ZOOM

Nature

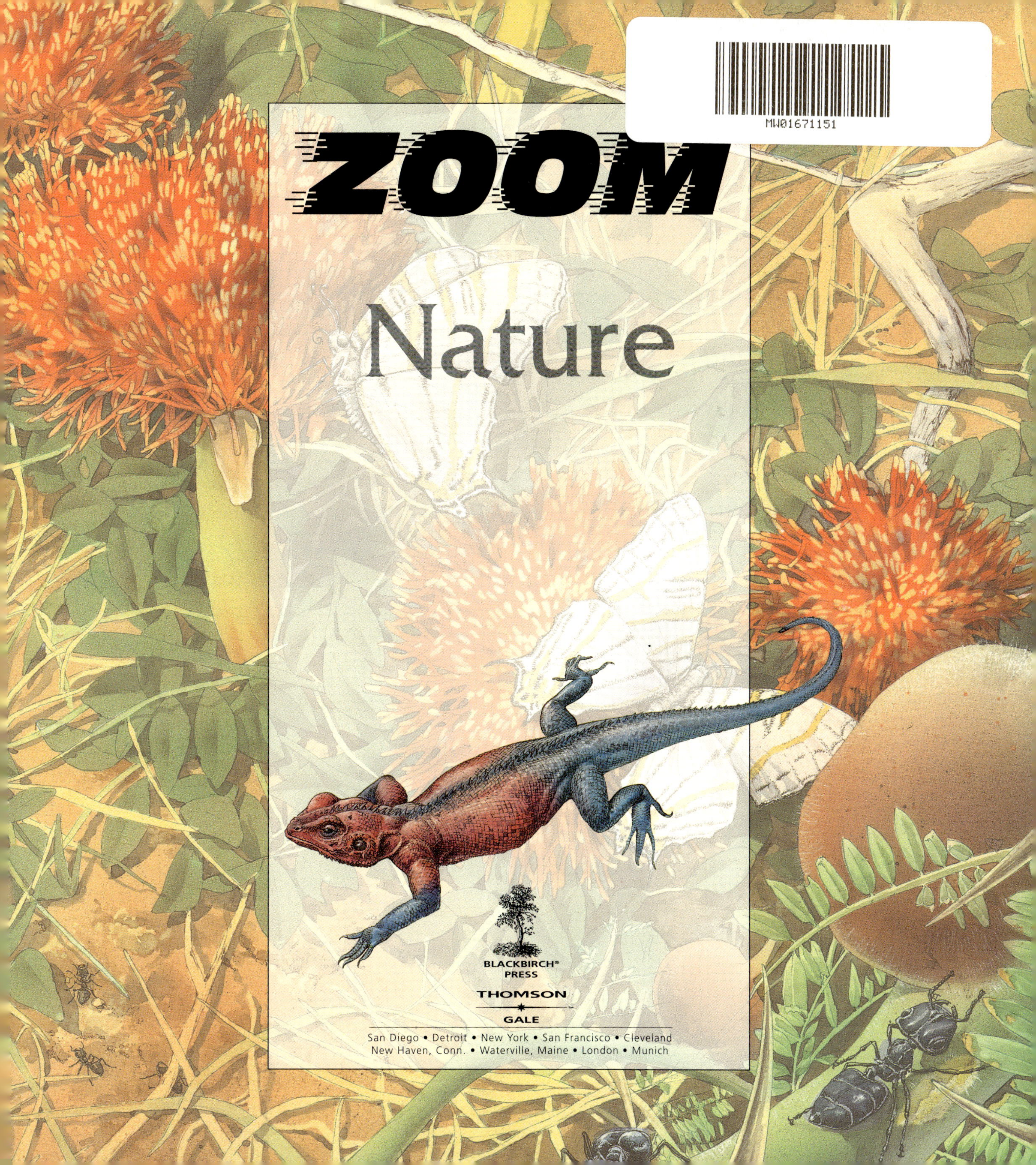

BLACKBIRCH® PRESS

THOMSON
GALE

San Diego • Detroit • New York • San Francisco • Cleveland
New Haven, Conn. • Waterville, Maine • London • Munich

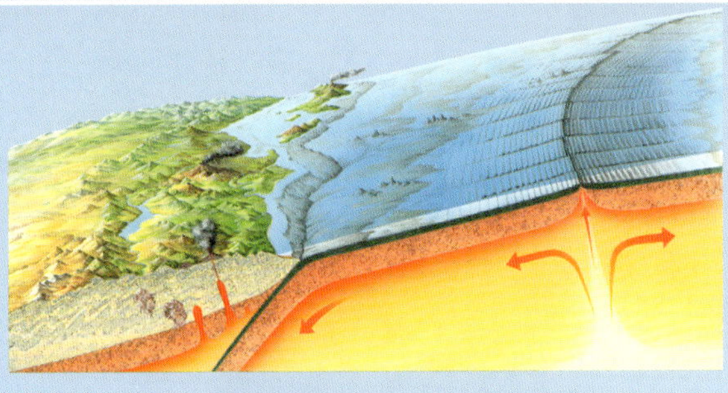

CONTENTS

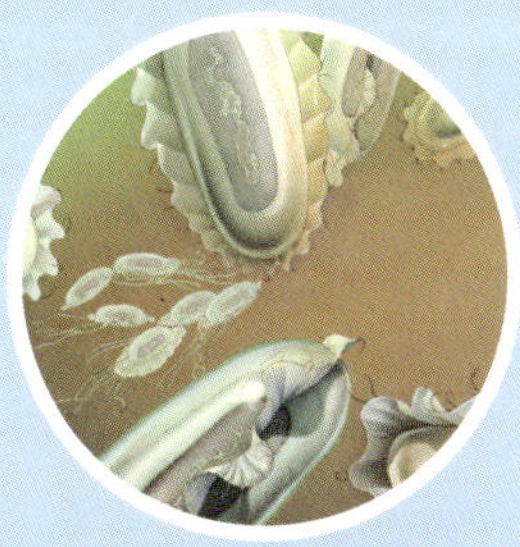

NATURE FACTFILE

LET'S ZOOM!

When you use the zoom feature on a camera, you bring an image from the distance into close-up focus without moving yourself. For example, you can capture the image of a butterfly on a leaf while keeping your distance. This book works in exactly the same way.

Imagine you were able to travel into space and point a camera back at the whole Earth. The illustration on pages 6 and 7 shows what you would see in your viewfinder. Now zoom in to one part of Earth's surface—the spectacular East African Rift Valley. You'll find yourself looking at a landscape of ancient volcanic calderas set amid grassy plains peppered with trees and bushes. Keep zooming, and you will begin to see some of the amazing inhabitants of this wild savanna landscape: elephants, giraffes, zebras, lions, and many other famous safari animals. Zoom in further and many much smaller and less familiar creatures of the undergrowth come into view. You'll discover the flowers hidden away in that undergrowth, ants creeping up a flower stalk, and—zooming into a new microscopic world—bacteria that live on an ant's body.

It's a fascinating journey, yet you will not have to move one inch! And you will discover some amazing things about life on Earth that only this incredible *zooming* book can show you . . .

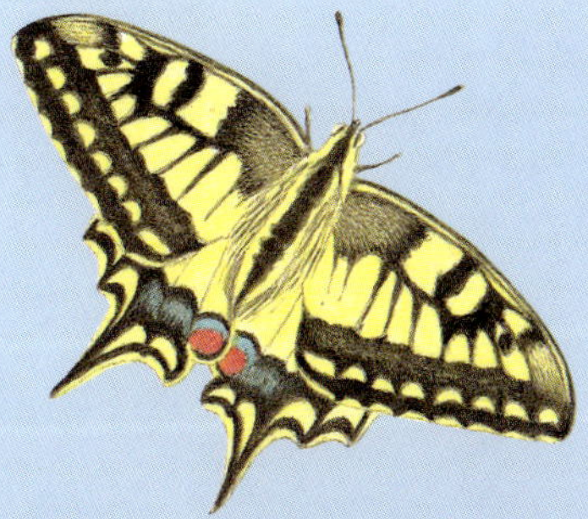

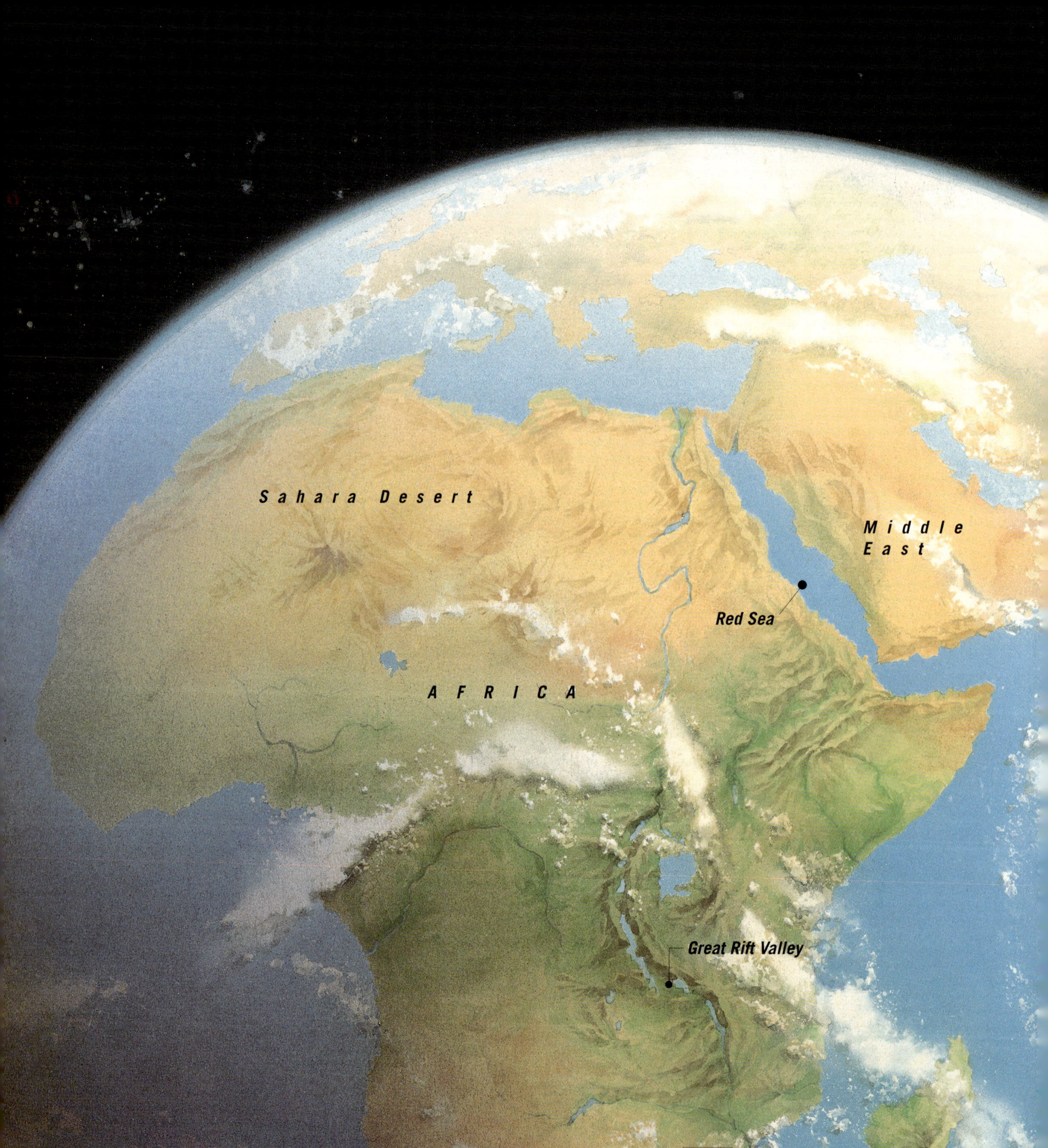
Sahara Desert
Middle East
Red Sea
AFRICA
Great Rift Valley

EARTH

Our planet is one of nine planets in the Solar System. It is the only known planet to have life, thanks to the presence of liquid water. Its atmosphere also supports life by shielding Earth from the Sun's harmful rays and from bombardment by large, rocky objects from space.

Earth is sometimes compared to a giant spaceship. It travels through space at more than 160,000 miles per hour (100,000 km/h), and all its inhabitants must depend on the planet's resources for survival. Essentials, such as water and nutrients, are recycled in nature, for the benefit of all Earth's living things *(see pages 23 and 26)*.

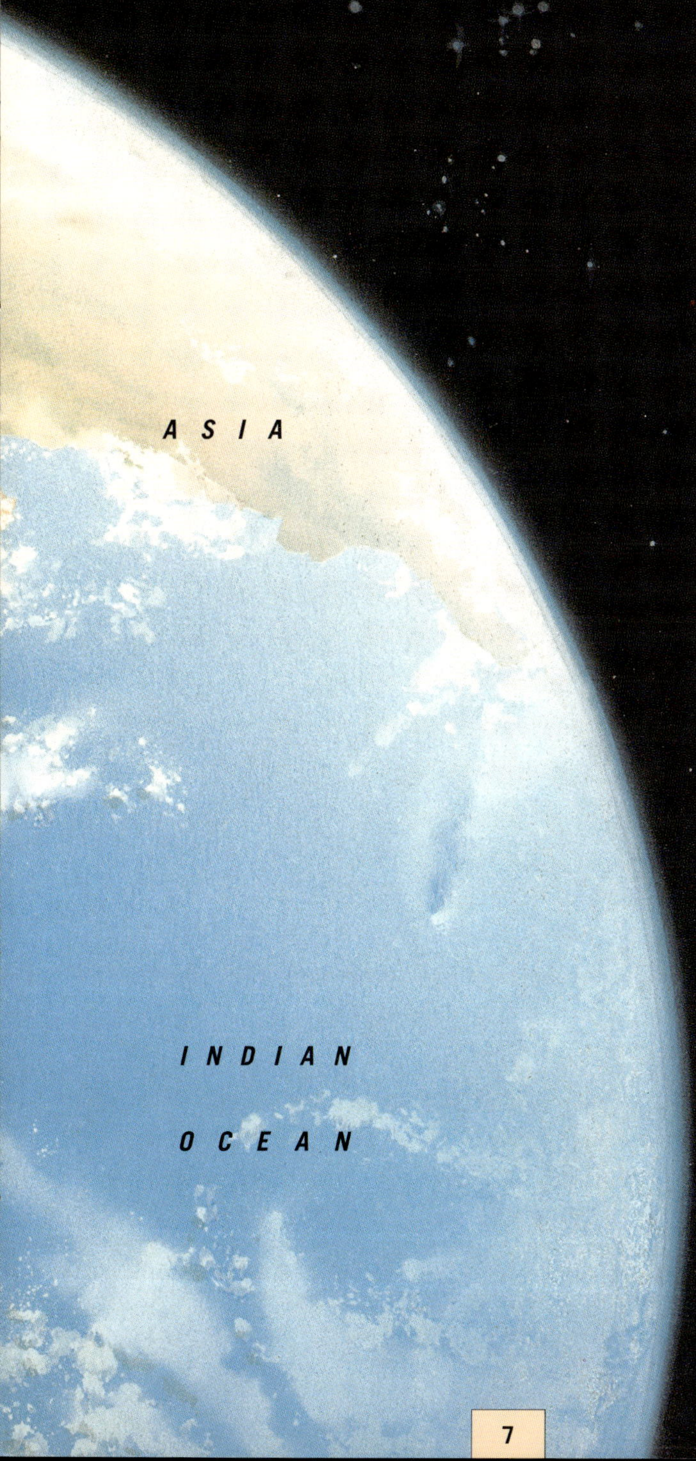

ZOOM DOWN TO THE LAND IN AFRICA, IN THE GREAT RIFT VALLEY

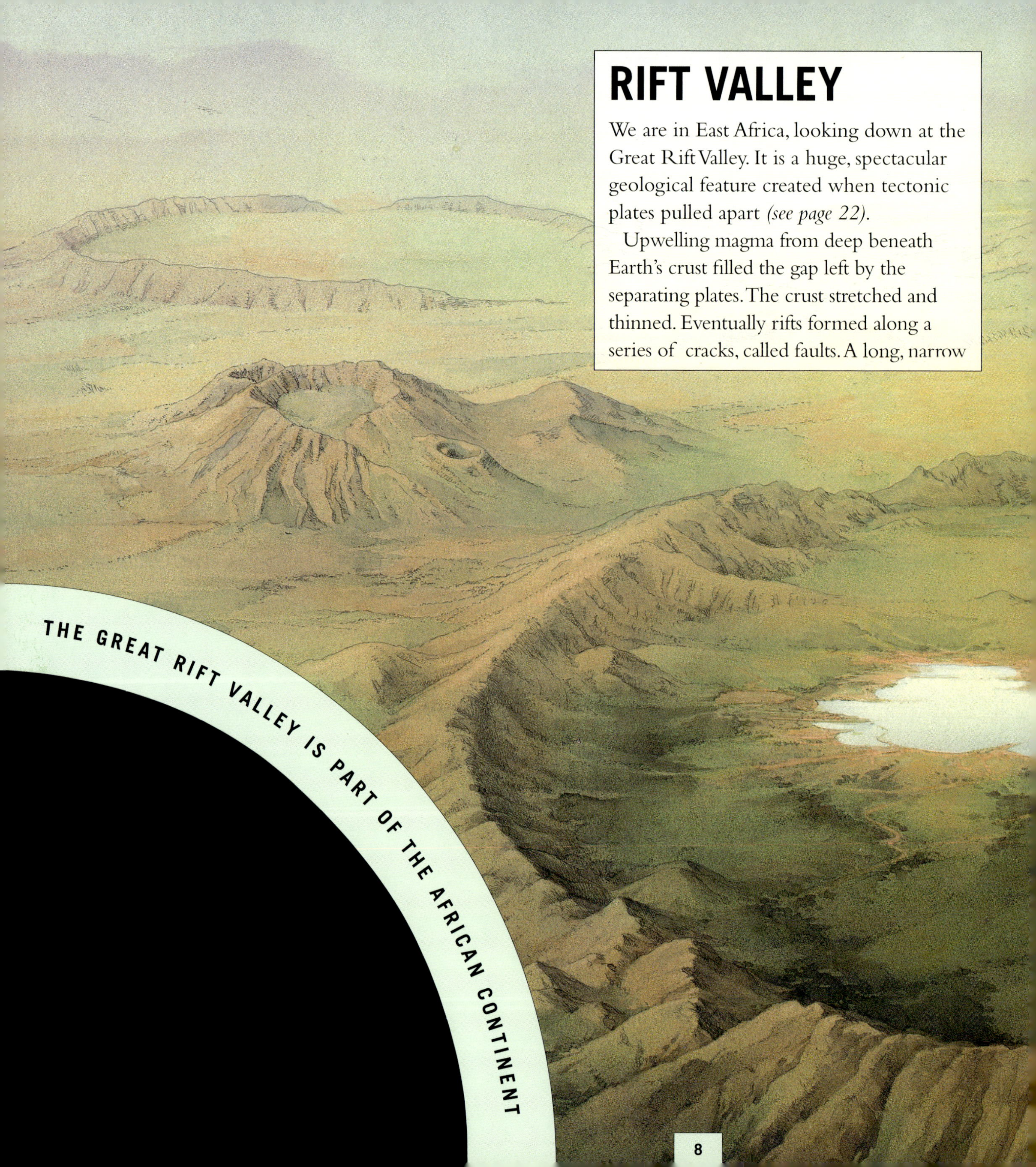

RIFT VALLEY

We are in East Africa, looking down at the Great Rift Valley. It is a huge, spectacular geological feature created when tectonic plates pulled apart *(see page 22)*.

Upwelling magma from deep beneath Earth's crust filled the gap left by the separating plates. The crust stretched and thinned. Eventually rifts formed along a series of cracks, called faults. A long, narrow

THE GREAT RIFT VALLEY IS PART OF THE AFRICAN CONTINENT

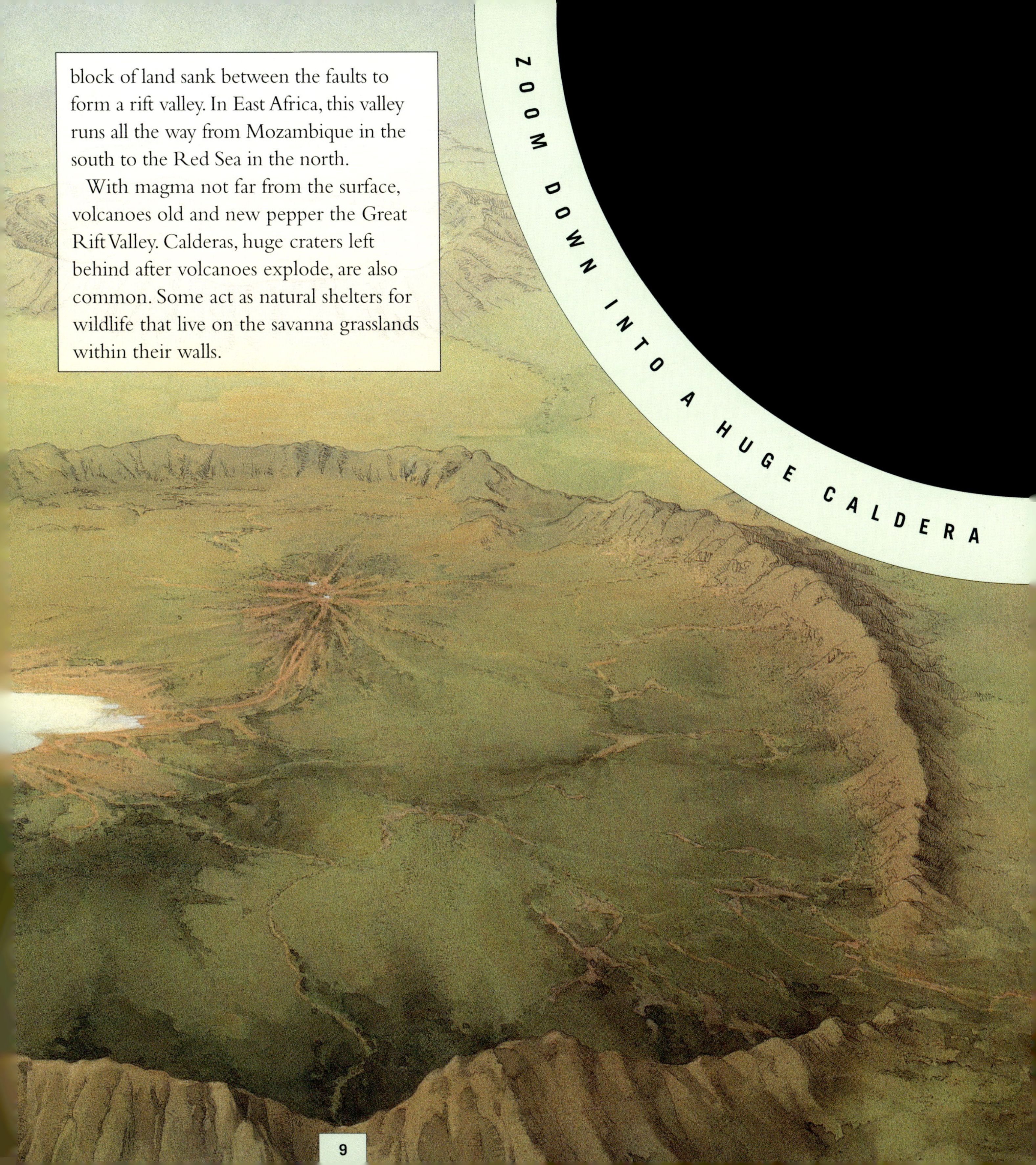

block of land sank between the faults to form a rift valley. In East Africa, this valley runs all the way from Mozambique in the south to the Red Sea in the north.

With magma not far from the surface, volcanoes old and new pepper the Great Rift Valley. Calderas, huge craters left behind after volcanoes explode, are also common. Some act as natural shelters for wildlife that live on the savanna grasslands within their walls.

SHIFTING TECTONIC PLATES CREATED THE RIFT VALLEY LANDSCAPE

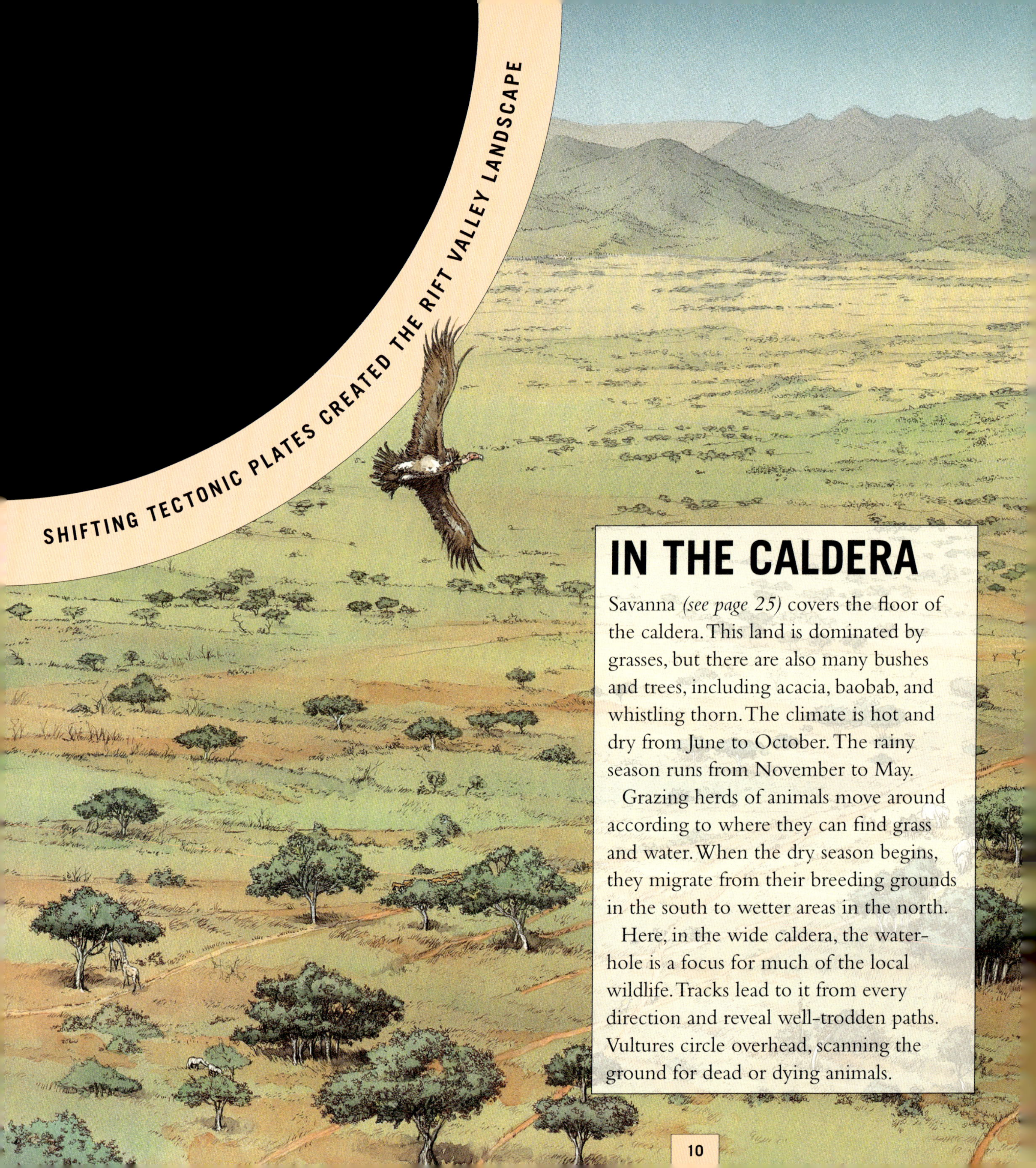

IN THE CALDERA

Savanna *(see page 25)* covers the floor of the caldera. This land is dominated by grasses, but there are also many bushes and trees, including acacia, baobab, and whistling thorn. The climate is hot and dry from June to October. The rainy season runs from November to May.

Grazing herds of animals move around according to where they can find grass and water. When the dry season begins, they migrate from their breeding grounds in the south to wetter areas in the north.

Here, in the wide caldera, the water-hole is a focus for much of the local wildlife. Tracks lead to it from every direction and reveal well-trodden paths. Vultures circle overhead, scanning the ground for dead or dying animals.

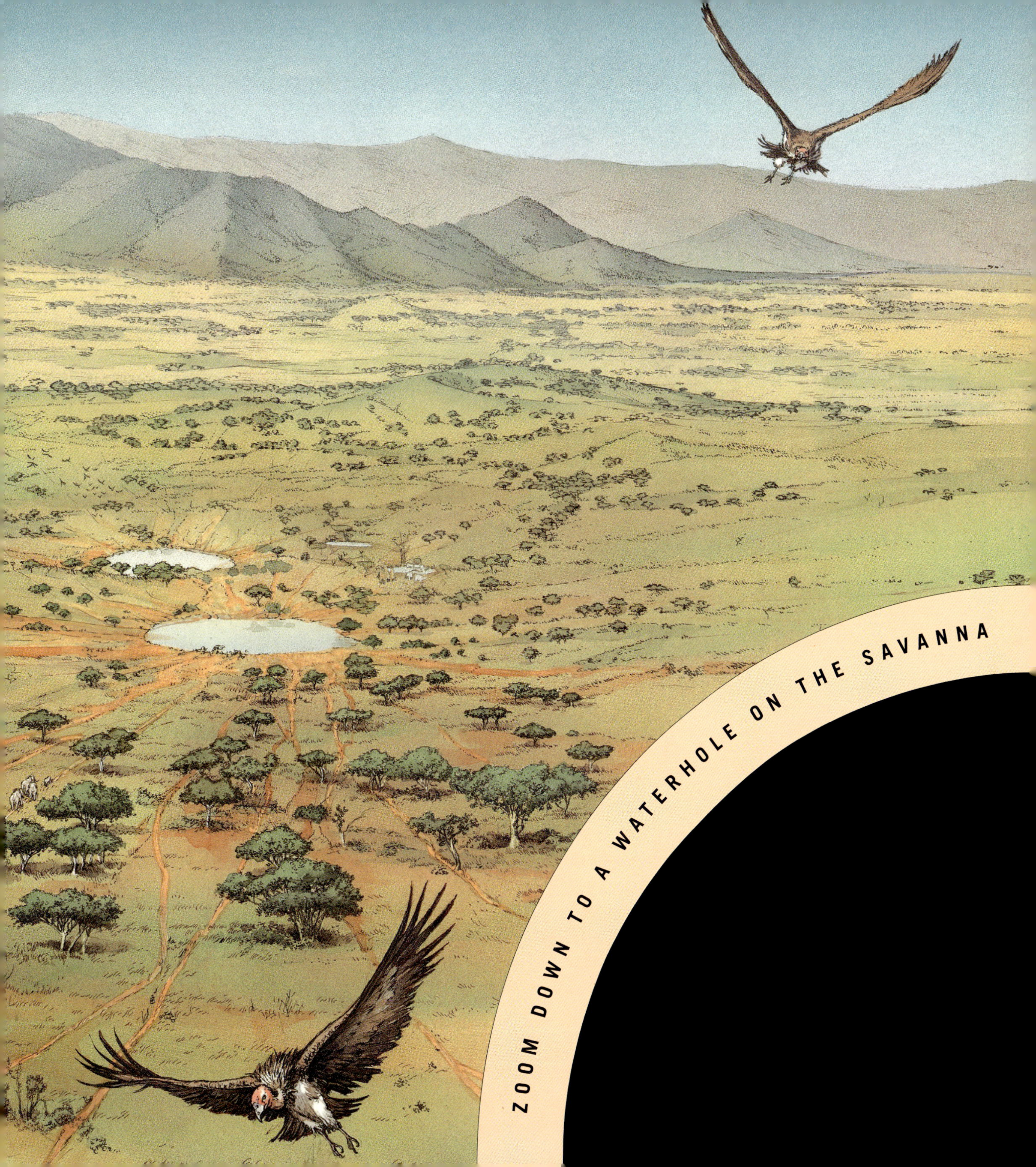
ZOOM DOWN TO A WATERHOLE ON THE SAVANNA

ANIMALS

The savanna grasslands of East Africa are home to a wide range of plant-eating animals. The grazers—animals that feed on grass—include zebras, which eat the coarse tops, and gazelles, which crop the shoots close to the ground. Among the leaf eaters are giraffes, which eat from the topmost shoots of the trees, and elephants, which take leaves (and bark) from lower down. Rhinoceroses feed on small shrubs and plants.

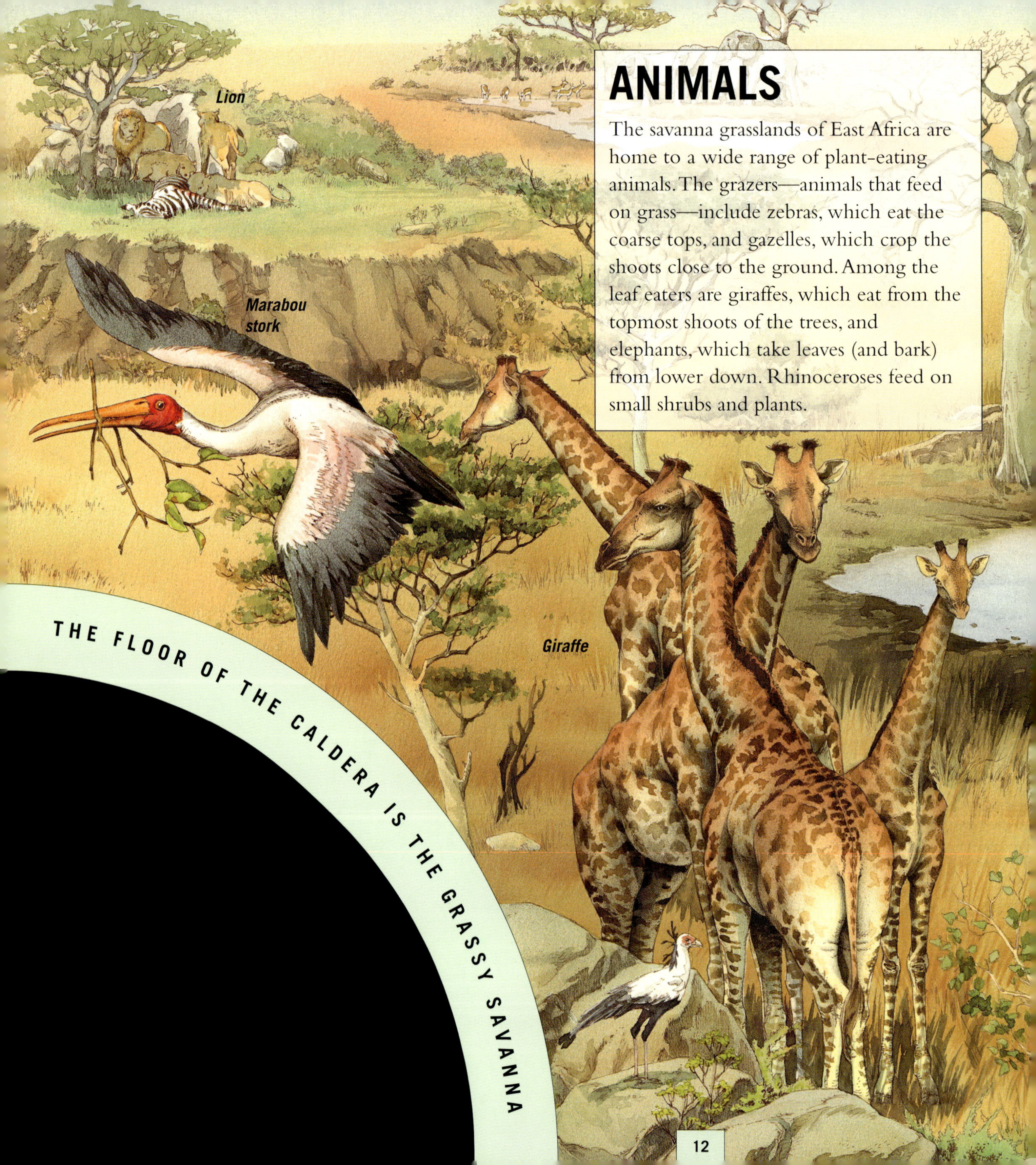

THE FLOOR OF THE CALDERA IS THE GRASSY SAVANNA

The savanna's meat-eating animals prey on some of the plant eaters. The lion catches its victims by making short bursts from the cover of long grass or bushes. Scavengers, including storks, vultures, and hyenas (which also make their own kills) feed on the carcasses left behind by lions and others.

ZOOM DEEP IN TO THE TALL SAVANNA GRASS

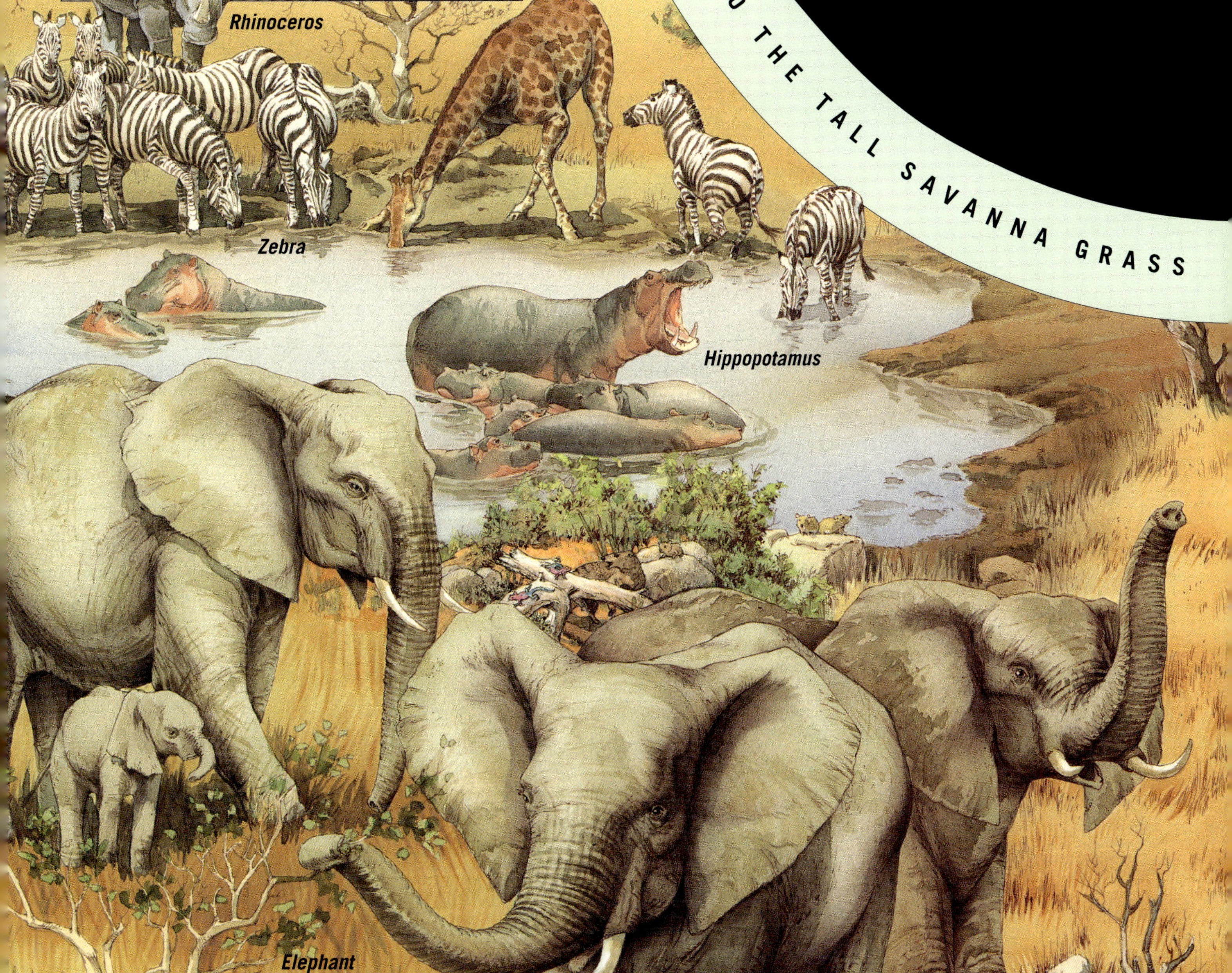

THE SAVANNA IS FILLED WITH GRAZERS, LEAF EATERS, AND MEAT EATERS

SMALL ANIMALS

Smaller creatures also inhabit the savanna. Banded mongooses, only 1.5 feet (0.5 m) long, hunt lizards, mice, insects, and other small prey. Mongoose packs may travel many miles a day in search of food during the dry season.

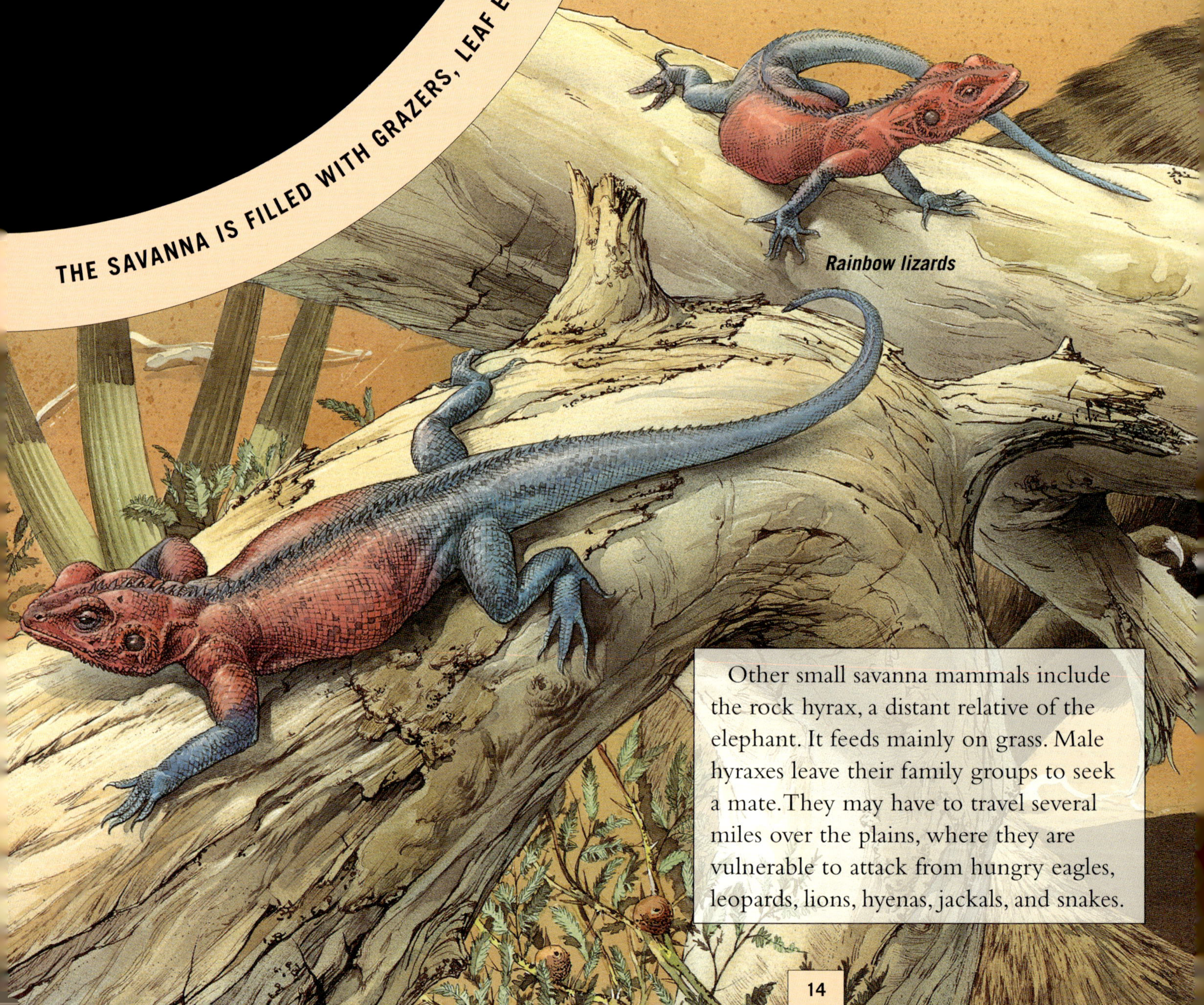

Rainbow lizards

Other small savanna mammals include the rock hyrax, a distant relative of the elephant. It feeds mainly on grass. Male hyraxes leave their family groups to seek a mate. They may have to travel several miles over the plains, where they are vulnerable to attack from hungry eagles, leopards, lions, hyenas, jackals, and snakes.

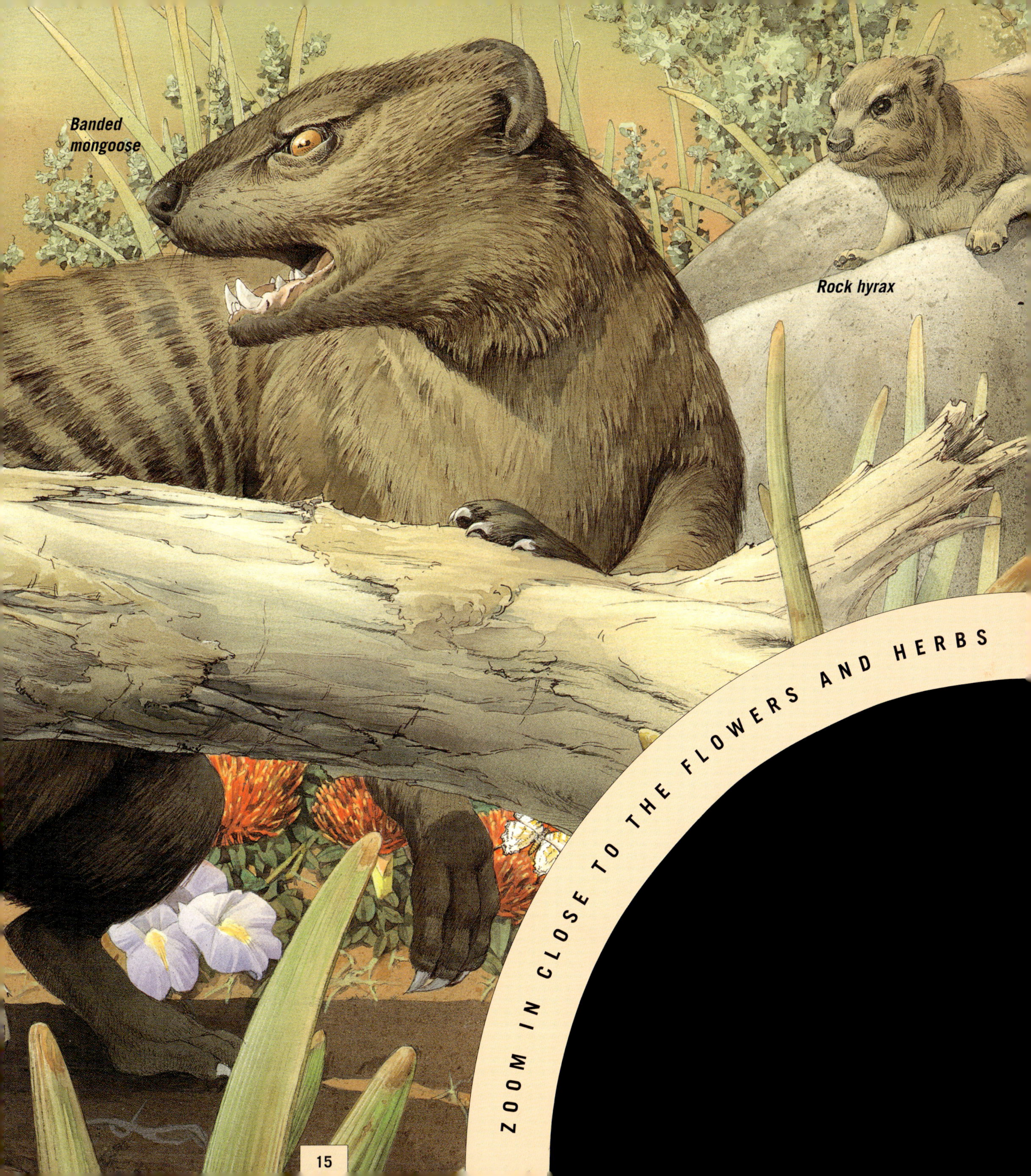

ZOOM IN CLOSE TO THE FLOWERS AND HERBS

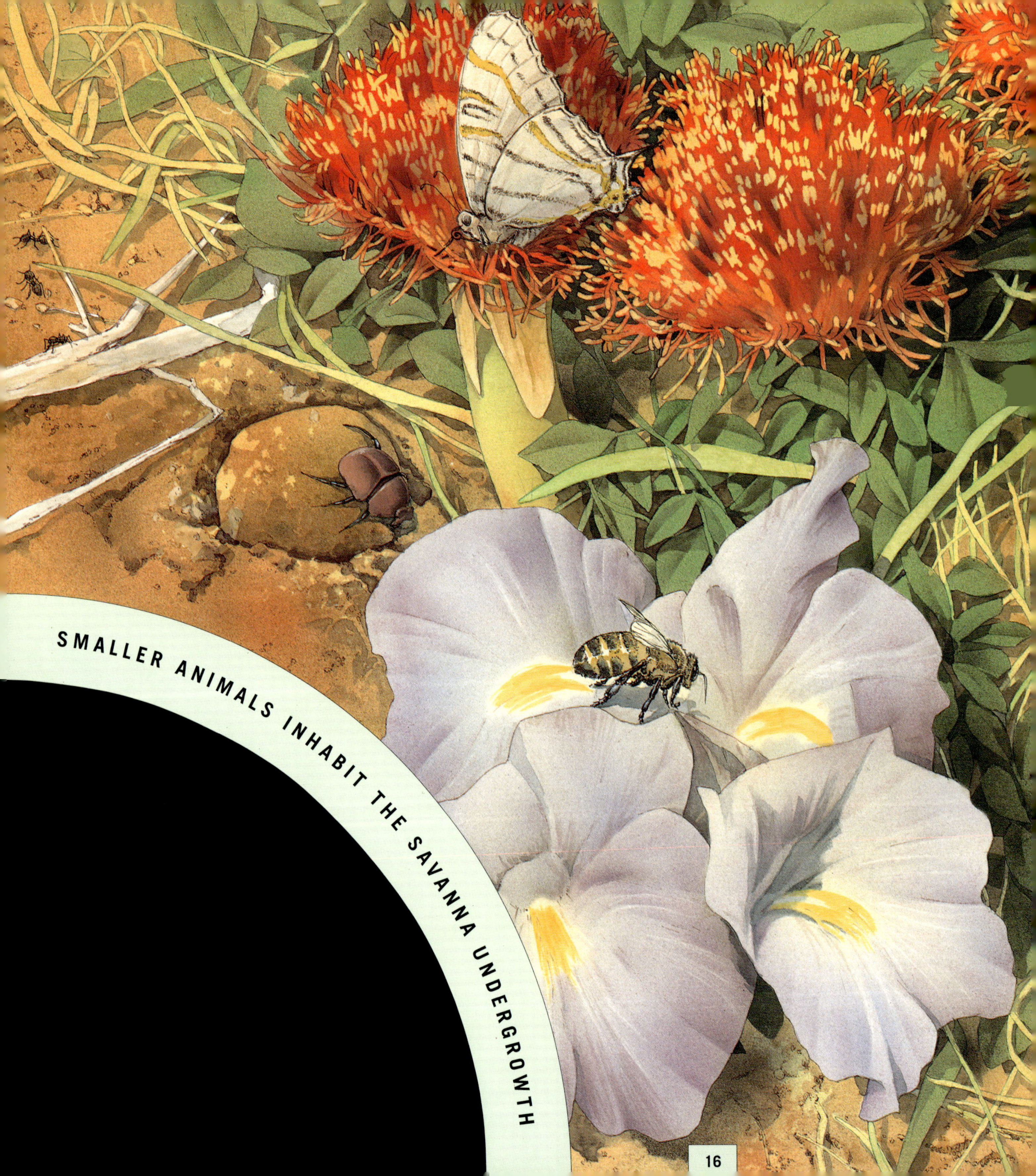
SMALLER ANIMALS INHABIT THE SAVANNA UNDERGROWTH

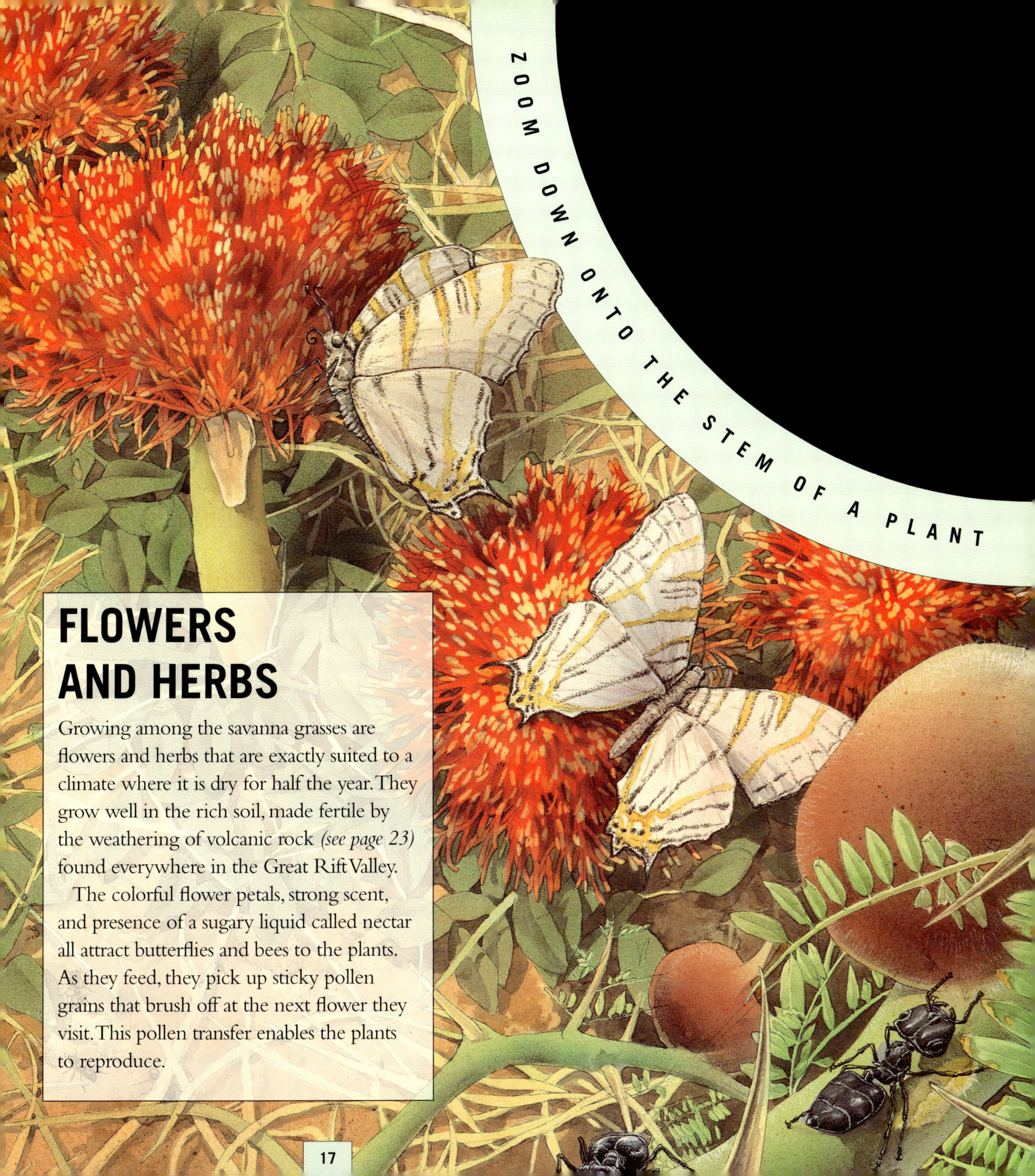

FLOWERS AND HERBS

Growing among the savanna grasses are flowers and herbs that are exactly suited to a climate where it is dry for half the year. They grow well in the rich soil, made fertile by the weathering of volcanic rock *(see page 23)* found everywhere in the Great Rift Valley.

The colorful flower petals, strong scent, and presence of a sugary liquid called nectar all attract butterflies and bees to the plants. As they feed, they pick up sticky pollen grains that brush off at the next flower they visit. This pollen transfer enables the plants to reproduce.

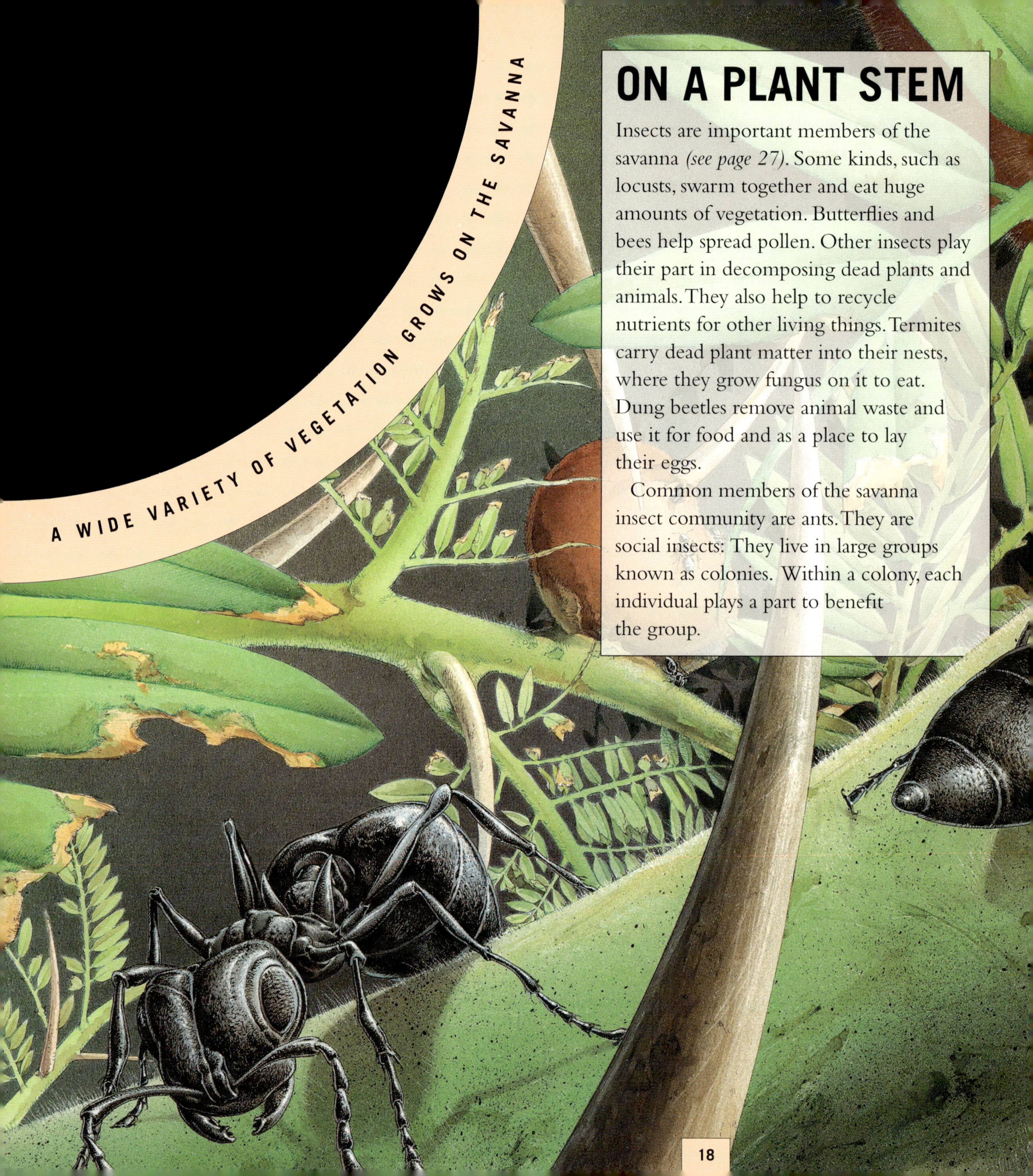

A WIDE VARIETY OF VEGETATION GROWS ON THE SAVANNA

ON A PLANT STEM

Insects are important members of the savanna *(see page 27)*. Some kinds, such as locusts, swarm together and eat huge amounts of vegetation. Butterflies and bees help spread pollen. Other insects play their part in decomposing dead plants and animals. They also help to recycle nutrients for other living things. Termites carry dead plant matter into their nests, where they grow fungus on it to eat. Dung beetles remove animal waste and use it for food and as a place to lay their eggs.

Common members of the savanna insect community are ants. They are social insects: They live in large groups known as colonies. Within a colony, each individual plays a part to benefit the group.

Worker ants are females that carry out a number of tasks. They tend to the eggs and larvae, and forage for food. The workers shown here are searching for food among the stems of plants in the undergrowth. They will then return to the nest to pass it on to the other ants.

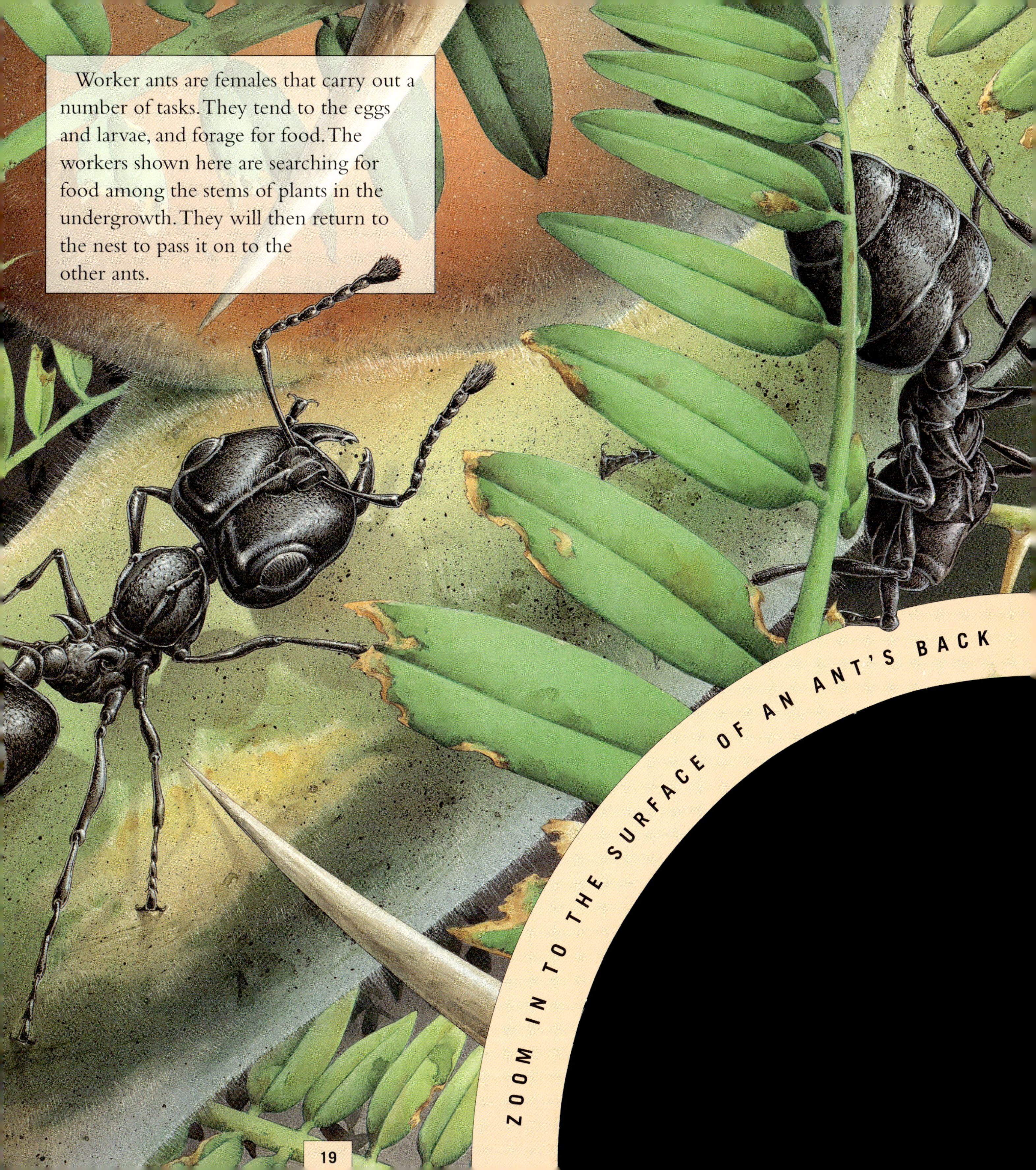

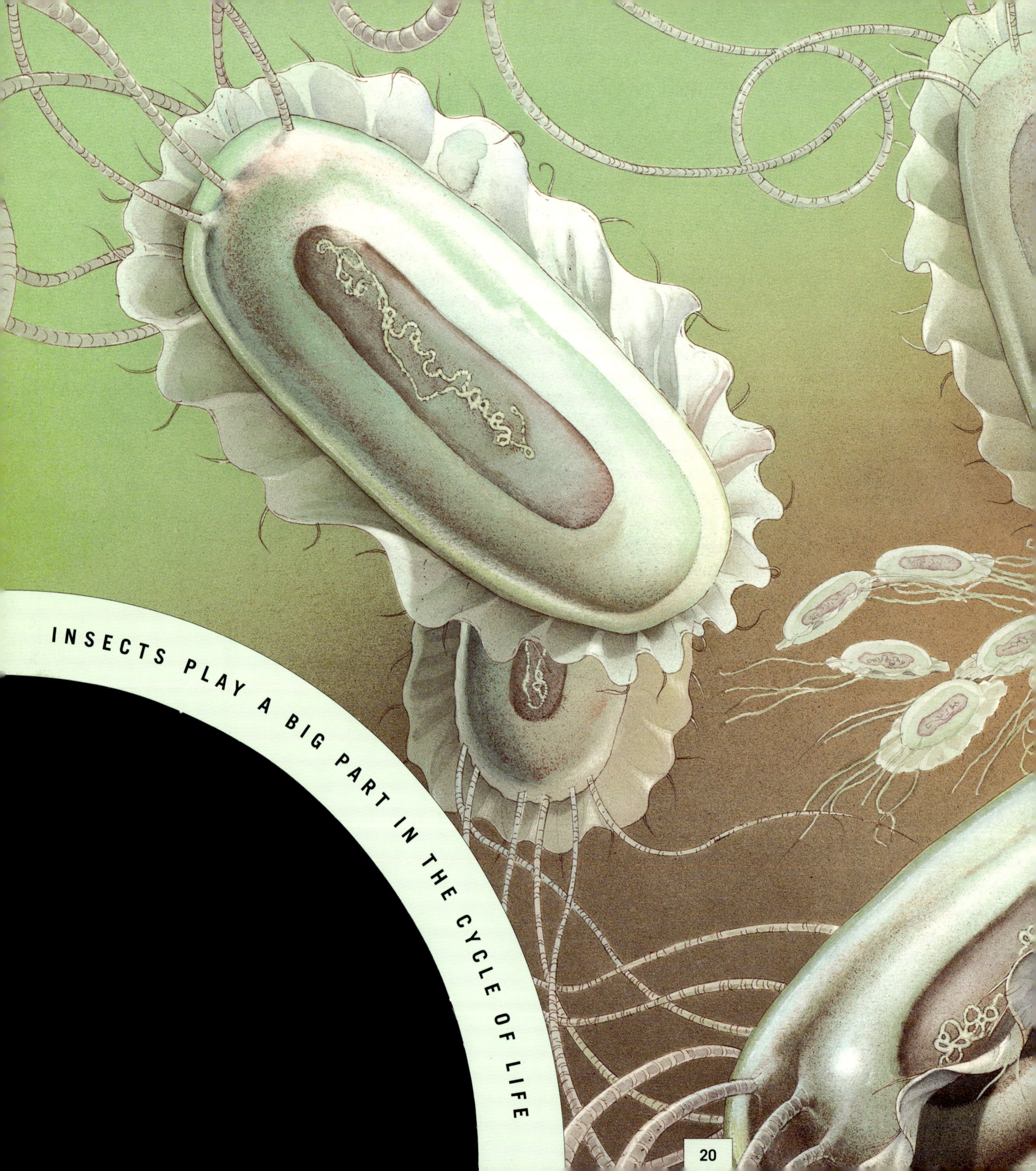

INSECTS PLAY A BIG PART IN THE CYCLE OF LIFE

ON AN ANT'S BACK

These bacteria live on the backs of ants.

Bacteria are all around us. They float in the air and drift in ocean depths. They live on icy mountaintops and inside rocks deep below Earth's surface. Some bacteria live inside humans.

These microscopic organisms range in size from 0.0002 to 0.05 millimeters. All bacteria have tough outer skins that contain a jelly-like substance called cytoplasm. Some have whip-like tails called flagella to help them move.

Certain bacteria are vital to life on Earth because they cause the decay of dead plants and animals, thus returning nutrients to the soil *(see page 26)*. Other bacteria may cause diseases when they get inside another living thing.

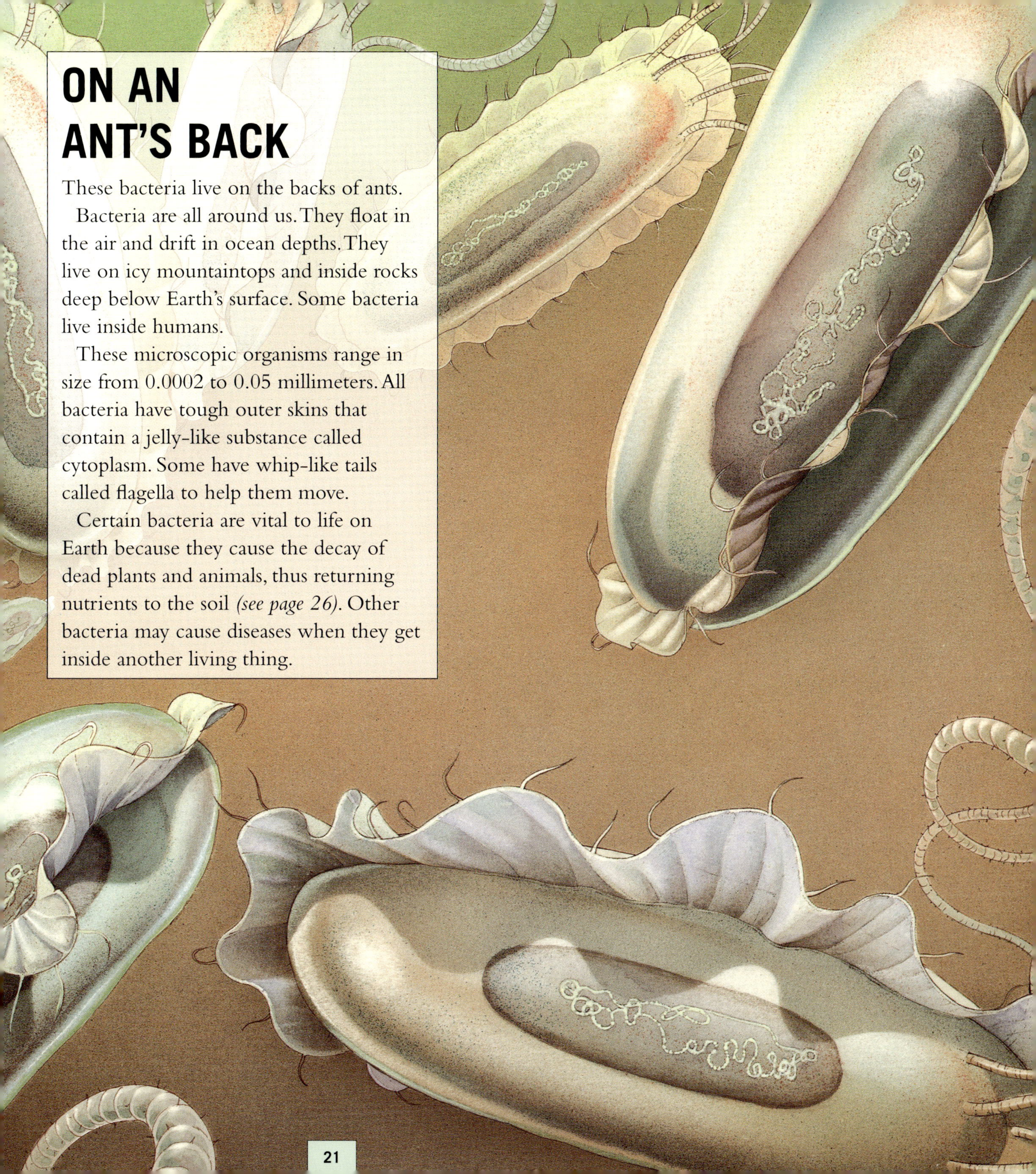

PLANET EARTH

Earth is made up of several layers *(right)*. The outer layer is a thin, rocky **crust** just a few miles thick. Beneath it lies the **mantle**. The mantle is divided into the upper mantle—made of a mix of molten and solid rock—and the lower mantle—a band of rock made solid by the pressure of the layers above. At the center lies the **core**, made mostly of iron. The outer core is liquid, while the inner core is solid. The temperature rises from 2,732°F (1,500°C) at the base of the crust to 13,530°F (7,500°C) at the center of Earth.

Earth's outer shell is divided into about 15 jagged-edged pieces called **tectonic plates**. In some places the edges collide and crumple up to form mountains. In others, the plates are pushed apart by magma (molten rock) as it rises from the mantle. Plates slide beneath one another near the edges of some oceans *(below)* and can set off volcanic eruptions and earthquakes. Plates sliding past one another in the opposite direction (as along the San Andreas Fault in California) also cause earthquakes.

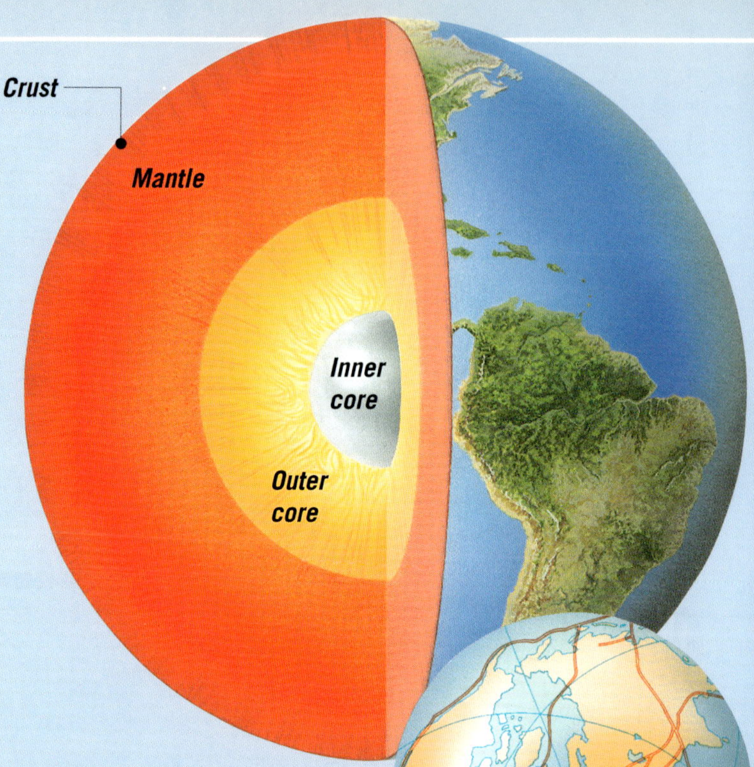

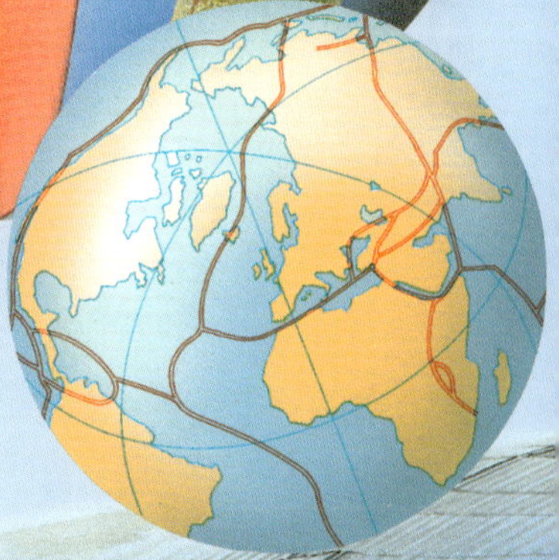

Earth's outer shell (the crust and the top of the mantle) is split into tectonic plates.

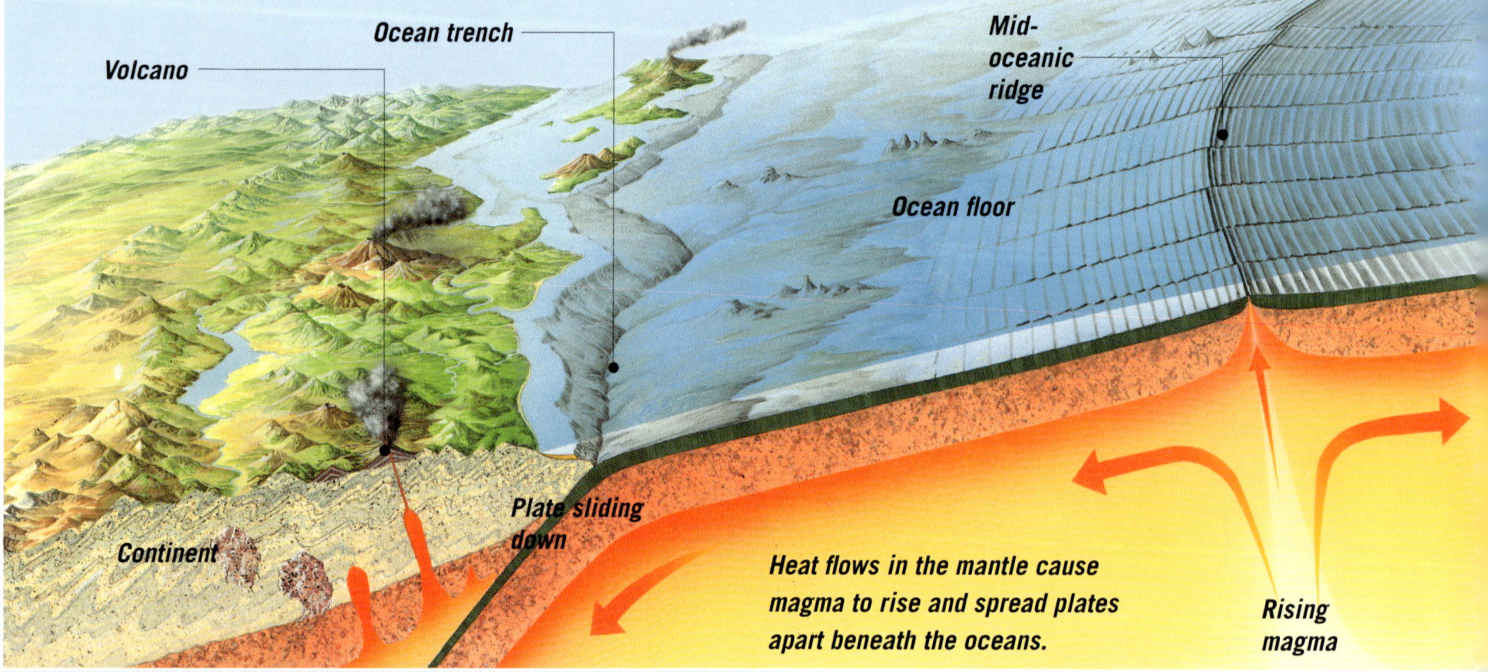

Heat flows in the mantle cause magma to rise and spread plates apart beneath the oceans.

Earth is the only planet to have liquid water. It circulates between its surface and the atmosphere—the blanket of air that surrounds the planet—in an endless cycle known as the **water cycle** *(right)*.

The water cycle plays a vital role in shaping Earth's surface. Though they seem hard and permanent, all rocks are vulnerable to **weathering**. Sharp changes in temperature crack rocks open. Rainwater in rock crevices expands as it freezes, splitting off pieces. Constant battering by waves, rivers, or slow grinding by glaciers also wears away rocks. Over millions of years, wide valleys and the leveling of mountain ranges result from this process of **erosion**.

Sedimentary rocks, such as sandstone and limestone, are made from sand, gravel, mud, and the other fragments that result from erosion (or from the remains of living things). As layers build up, they are compressed into rock. **Igneous rocks**, such as granite and basalt, are formed when magma rises, cools, and solidifies in Earth's crust. **Metamorphic rocks**, such as marble and slate, are formed when rocks are subjected to great pressure and heat. The three different types change from one to another in the **rock cycle** *(below)*.

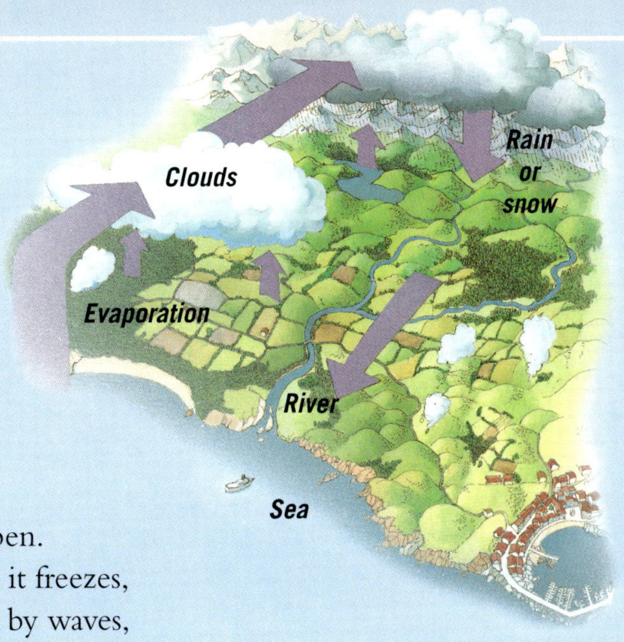

Warmed by the sun, water evaporates from lakes, rivers, and the oceans. The moisture is carried upward where it cools and condenses (turns back into liquid water). Tiny water droplets or ice crystals gather around dust particles to form clouds. Eventually they fall as rain or snow. Rivers carry the water back to the ocean.

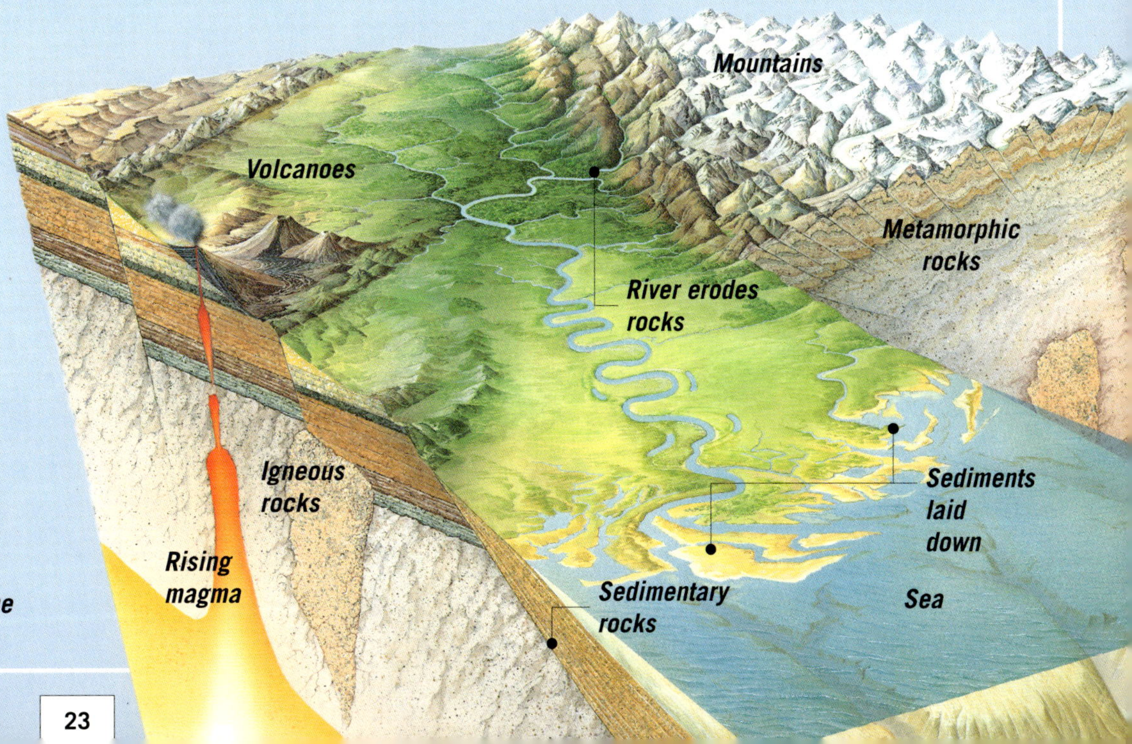

Sedimentary rocks formed under the sea may sink into Earth where one tectonic plate slides under another **(opposite)**. ***They may then melt, rise, and cool to become igneous rocks. Sedimentary rocks may also be crushed or cooked under the ground to form metamorphic rocks. All types of rock may be eroded, becoming sedimentary rocks again. This continual process is known as the rock cycle*** **(right)**.

CLIMATES AND BIOMES

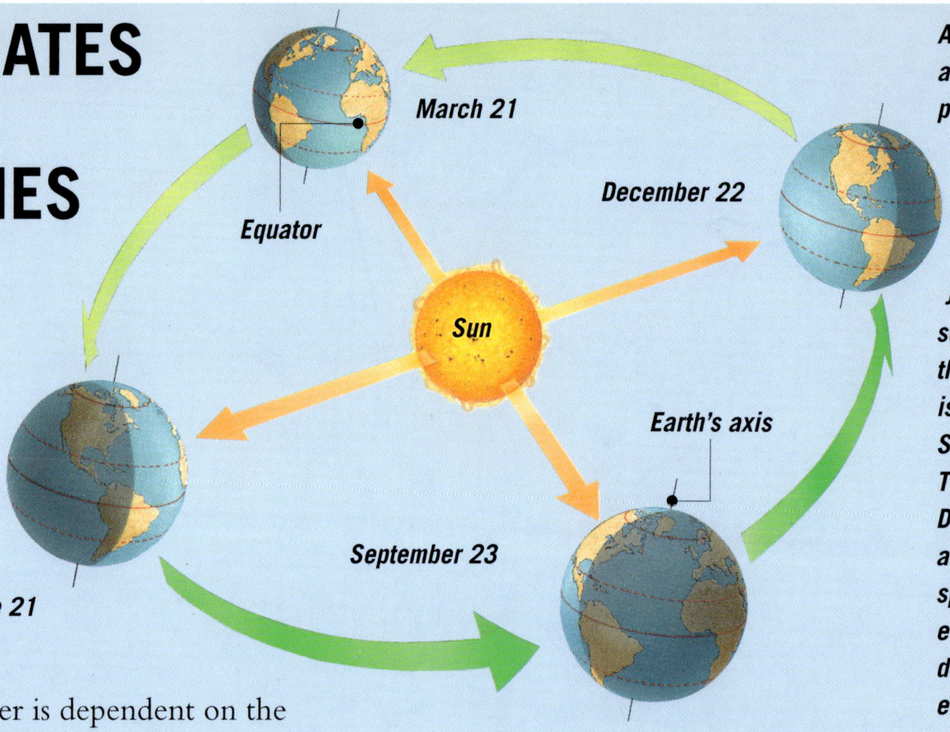

As Earth spins on its orbit around the Sun, it is not perfectly upright, but slightly tilted. This gives us the seasons. When Earth's northern hemisphere leans nearer the Sun on June 21, it is the summer solstice. At the same time, the southern hemisphere is tilted away from the Sun, so it is winter there. The opposite occurs on December 22. On March 21 and September 23, the spring and autumn equinoxes, the Sun shines directly overhead at the equator.

Our weather is dependent on the warmth from the Sun. Winds result from uneven warming of different parts of the world: Cool air blows in to replace rising warmer air. Rain and snow occur when water evaporated from the surface rises, cools, and condenses *(see page 23)*. Weather changes from day to day, but, over a longer period, regions of the world have a regular pattern of weather, called **climate**. Climates result from the way Earth orbits the Sun *(above)* and from the pattern of ocean currents and winds, which carry air and clouds around the globe.

Regions with tropical climates—hot, wet weather all year round—are at the equator where the Sun is closest to Earth. Tropical regions with dry winds have desert climates. At the poles the Sun is much lower in the sky, so it is much colder. Between the tropics and the poles are temperate lands that have warm summers and cool winters.

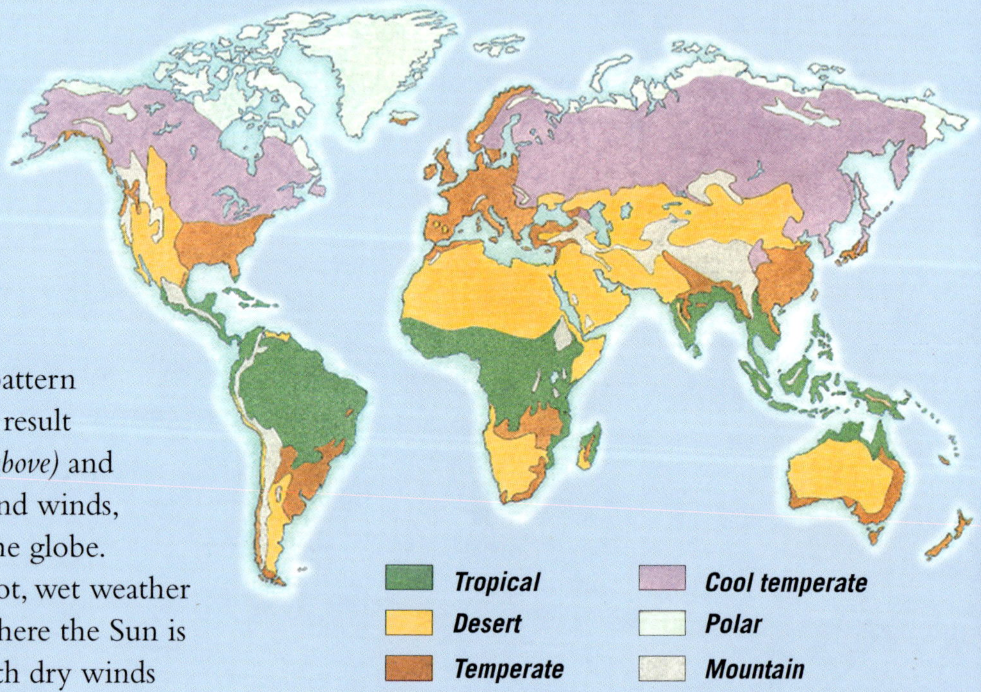

The pattern of world climates is partly influenced by ocean currents. For example, northwestern Europe is warmed by the flow of the Gulf Stream from the tropics.

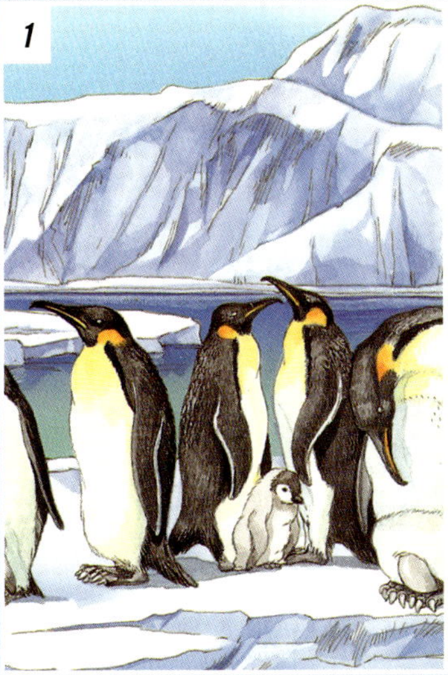
1

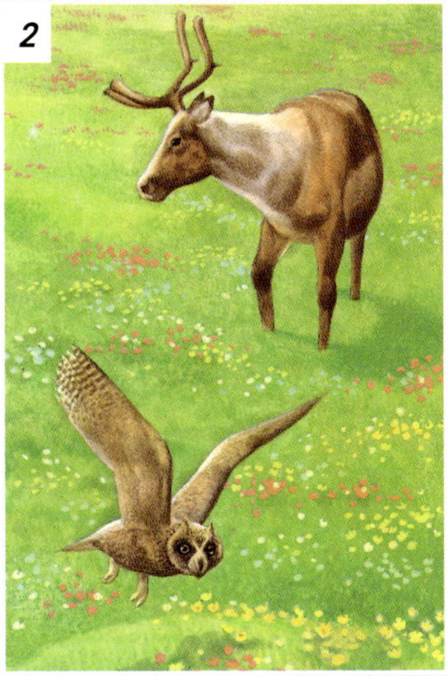
2

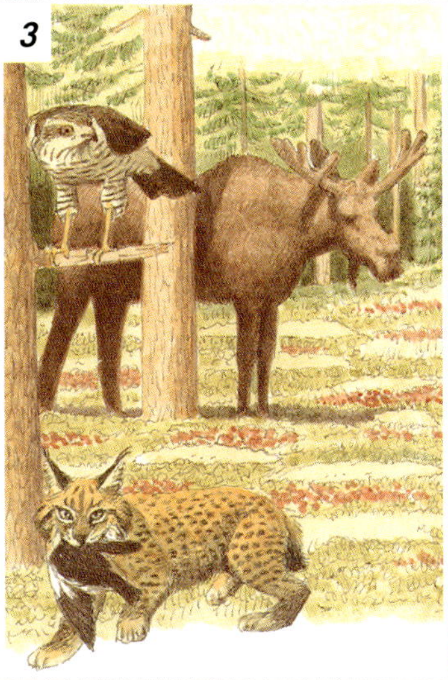
3

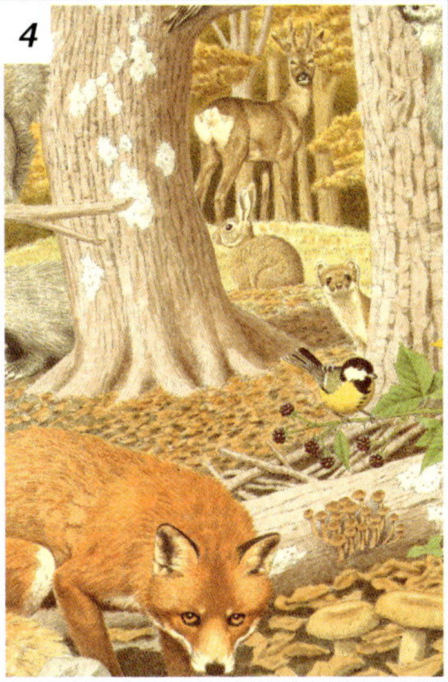
4

A **biome** includes all areas where the vegetation and animal life are broadly similar. Biomes are shaped by climate (and also the rocks and soil type). Each biome consists of a range of **habitats**. A temperate woodland biome, for example, may be made up of oak, beech, or maple habitats.

Earth's polar regions (1) are mostly covered by snow and ice. Close to northern polar lands is the tundra (2), a region too cold for trees, but where the snow melts in summer to let mosses grow. Boreal coniferous forest (3) covers much of northern Russia and North America, while deciduous woodland (4) grows in warmer climates

Most deserts (5) are bare and rocky. A few are sandy. Savanna grasslands (6) are hot and dry but also have a rainy season *(see page 10)*. Tropical rain forest (7) has high rainfall all year round. It is home to the richest variety of living things. The largest biome is the ocean (8). Most ocean life lives on or near the seabed in shallow waters.

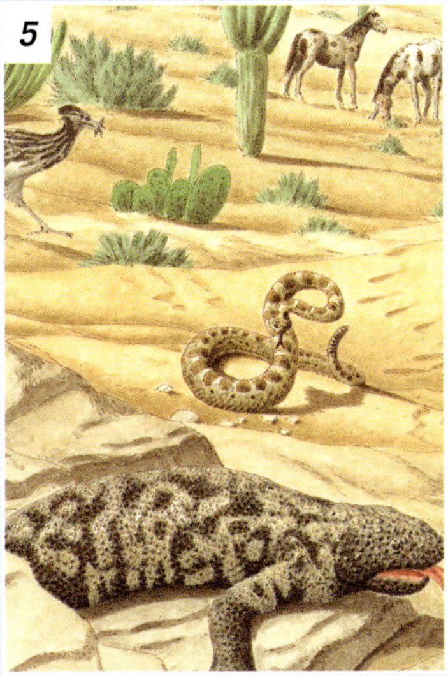
5

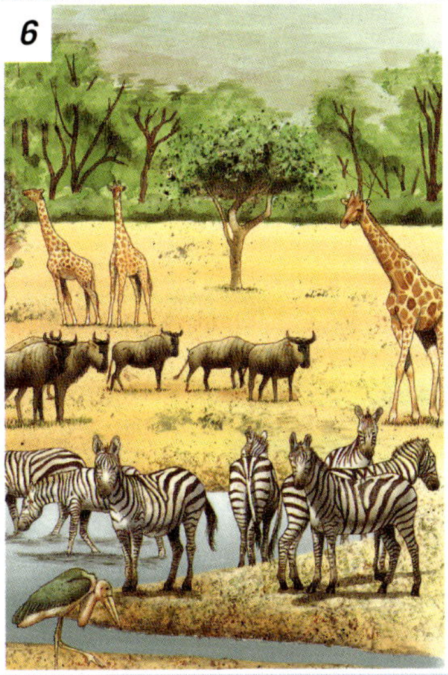
6

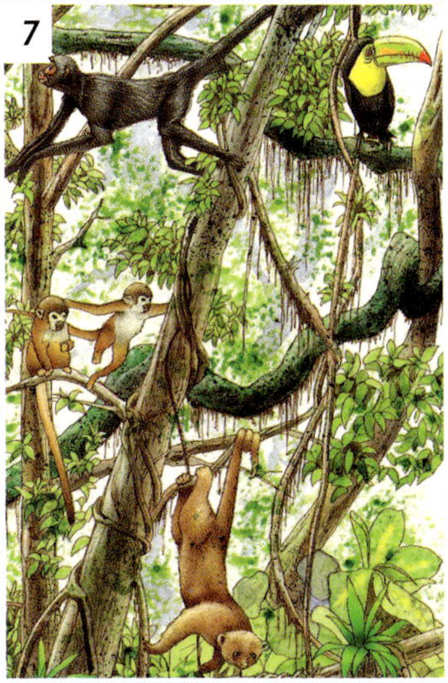
7

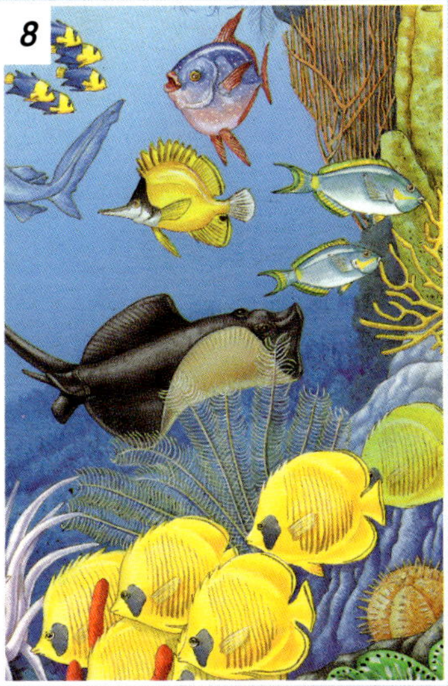
8

NATURAL CYCLES

The plants and animals of any habitat require **nutrients**—minerals and other substances—to survive and grow. In nature, chemical elements—such as oxygen, carbon, and sulfur—that make up nutrients are recycled via plants and the soil *(below)*. In this way, they are always available.

Plants suck up nutrients from the soil through their roots (1). They use these nutrients for nourishment, along with carbon dioxide from the air and light energy from the Sun. Plant-eating animals digest the nutrients in their food (2). The nutrients may pass on to a meat-eating animal, or return to the soil when either animal dies (3). Meanwhile, carbon dioxide breathed out by animals returns to the air. Fungi, microbes, and soil-dwelling animals all help the rotting process by which living things decay and release the nutrients, to be taken up by plants once more (4).

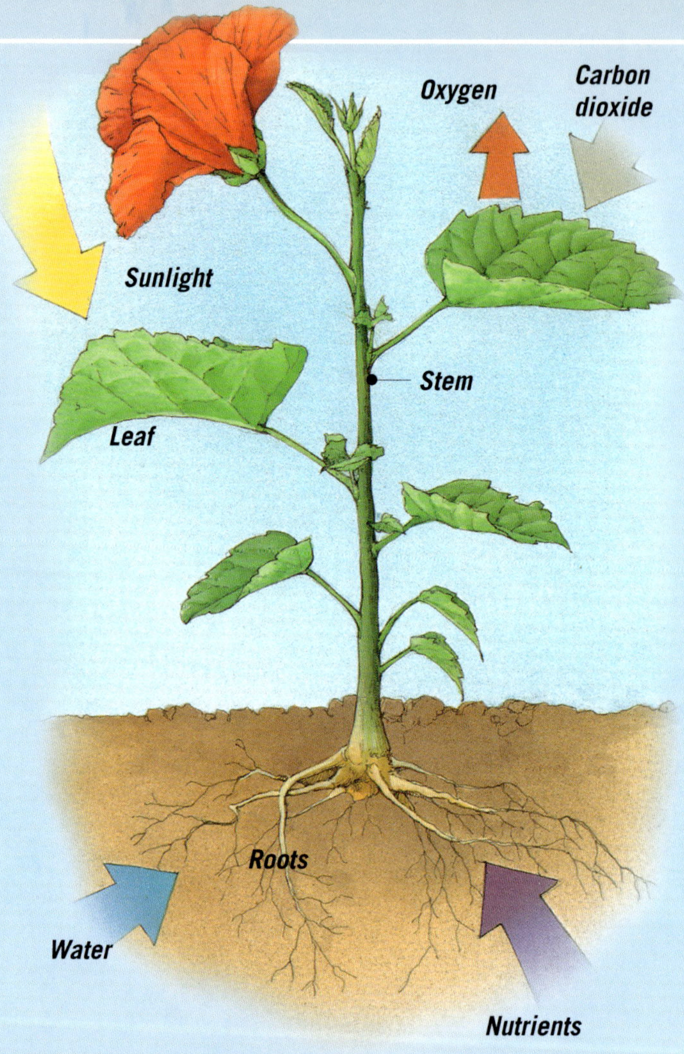

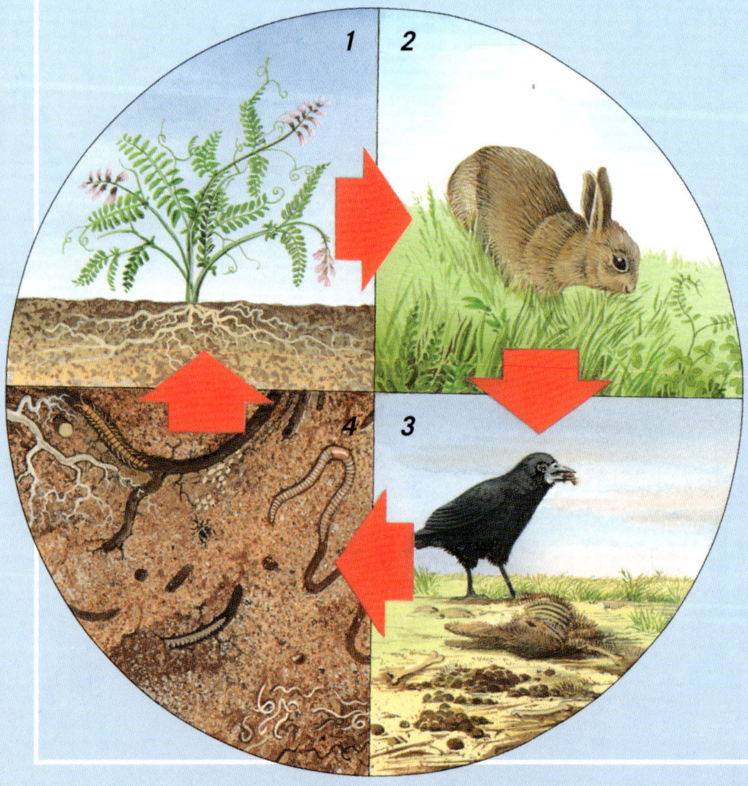

Plants play a vital role in the recycling of both nutrients and oxygen *(above)*. A plant's roots are the means for taking in nutrients and water from the soil. Its stem holds it above the surface and supports the leaves. A plant's leaves are its "food factory." A green substance called chlorophyll absorbs the energy in sunlight, which the plant uses to make a chemical reaction. The chemical reaction causes water from the soil and carbon dioxide from the air join together to form sugar, the plant's food. This process is called **photosynthesis**.

A byproduct of photosynthesis is oxygen, which seeps out into the air and is breathed in by animals.

Living things in any habitat interact and rely on each other for survival. Together, they form a **community**. Plants feed on sunlight, water, and nutrients from the soil on land, or from the water around them in the ocean. Animals known as **herbivores** in turn feed on plants. They range from tiny insects to mammals and birds. Animals that eat animals are called **carnivores**. The linking of plants and animals in this way is called a **food chain**. In nature, animals eat more than one kind of food—for example, a bear may eat herbivores, insect eaters, and insects themselves—so food chains are part of more complicated **food webs**, such as this one *(right)* from a North American forest. If one animal or plant in a food chain disappears for any reason, it can have disastrous effects on the food web.

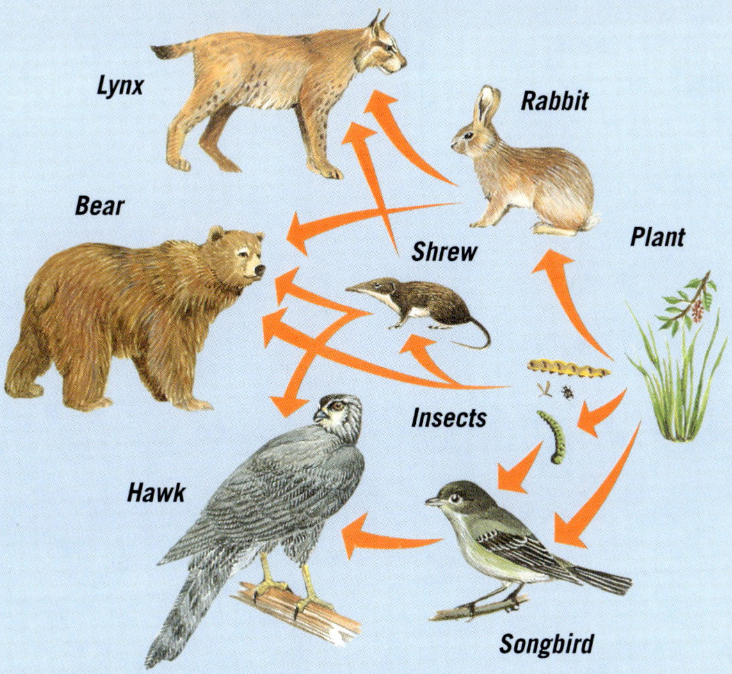

An individual living thing in a community is called an **organism**. To understand how organisms live, it helps to classify them or put them into groups, known as kingdoms: Monerans are the smallest, simplest organisms; protists are slightly more complex; fungi, plants, and animals are others *(see page 28)*.

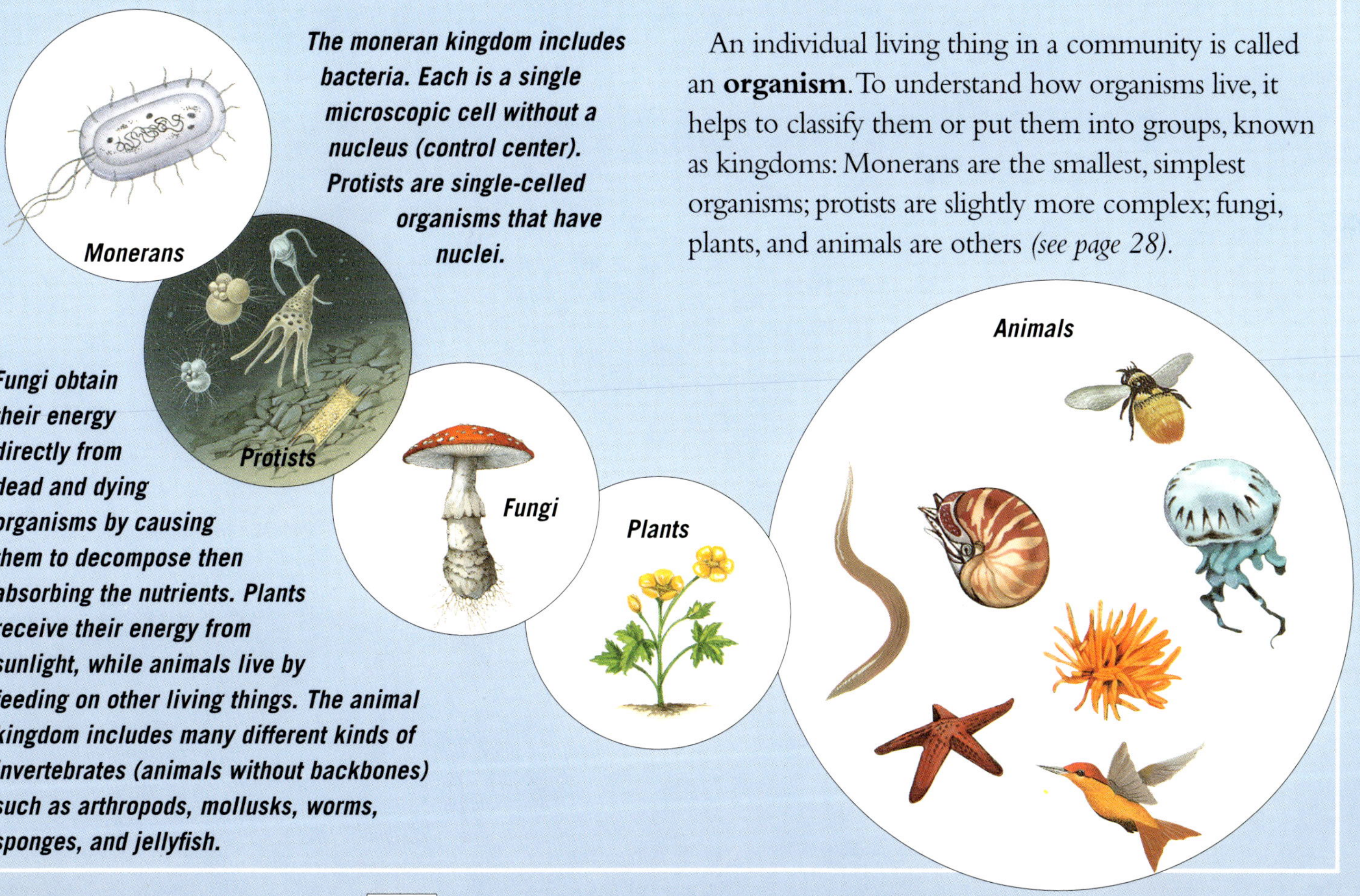

The moneran kingdom includes bacteria. Each is a single microscopic cell without a nucleus (control center). Protists are single-celled organisms that have nuclei.

Fungi obtain their energy directly from dead and dying organisms by causing them to decompose then absorbing the nutrients. Plants receive their energy from sunlight, while animals live by feeding on other living things. The animal kingdom includes many different kinds of invertebrates (animals without backbones) such as arthropods, mollusks, worms, sponges, and jellyfish.

ANIMAL GROUPS

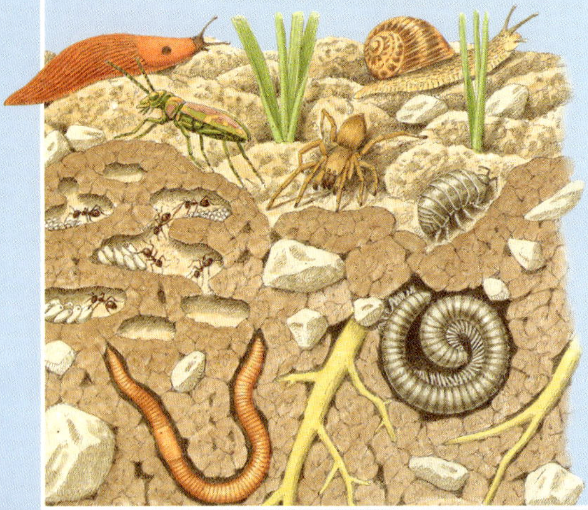

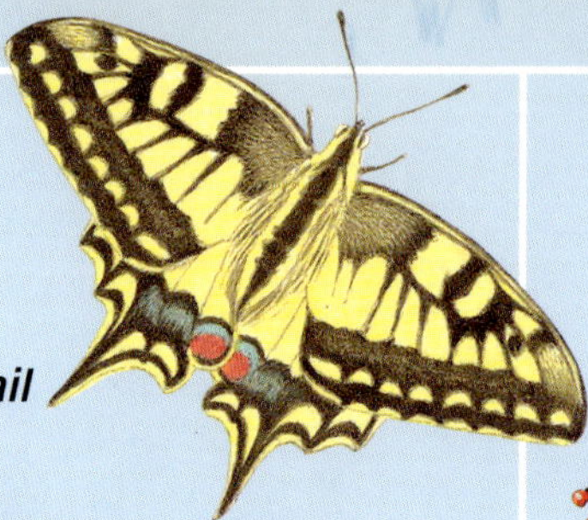

Swallowtail butterfly

Invertebrates: animals without backbones. This group includes many marine creatures (sponges, jellyfish, sea anemones, mollusks and crustaceans), microorganisms (protozoa), insects, spiders, millipedes, and centipedes.

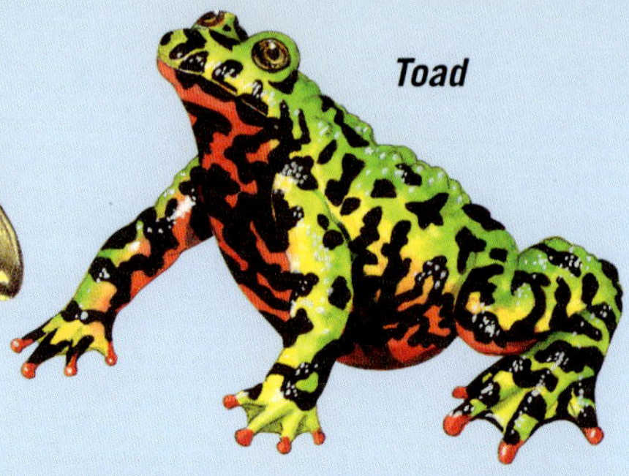

Toad

Amphibians: animals that return to water to breed. They include newts, salamanders, frogs, and toads. The young are aquatic and breathe using gills. Adults develop lungs.

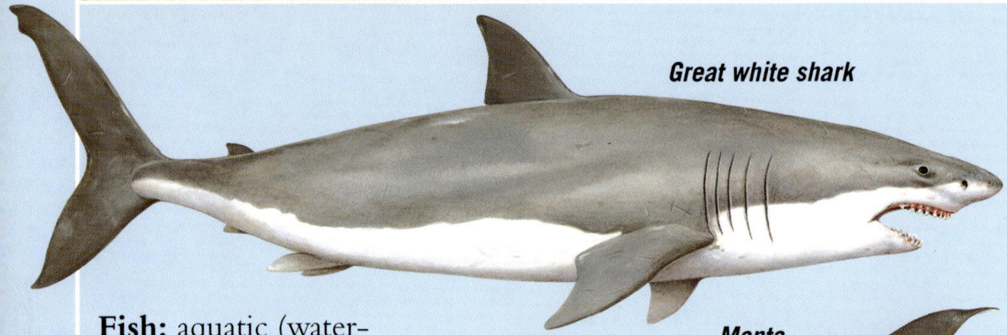

Great white shark

Fish: aquatic (water-living) animals that breathe by means of gills. They are cold-blooded, hairless, and streamlined in shape. Many have a protective layer of overlapping scales. Their fins help them to balance and steer.

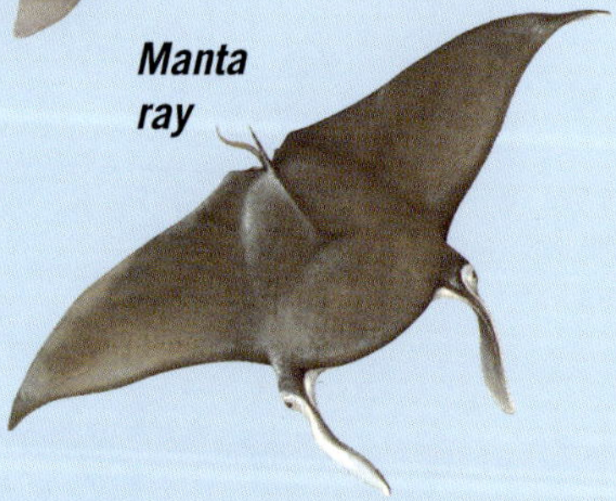

Manta ray

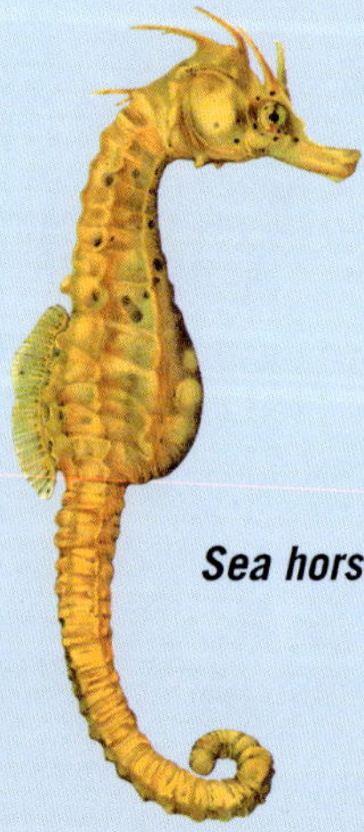

Sea horse

There are two main fish groups: cartilaginous fish (skeletons made of soft cartilage: sharks, rays, and chimeras) and bony fish (bony skeletons: sea horses, salmon, cod, eels, and others.)

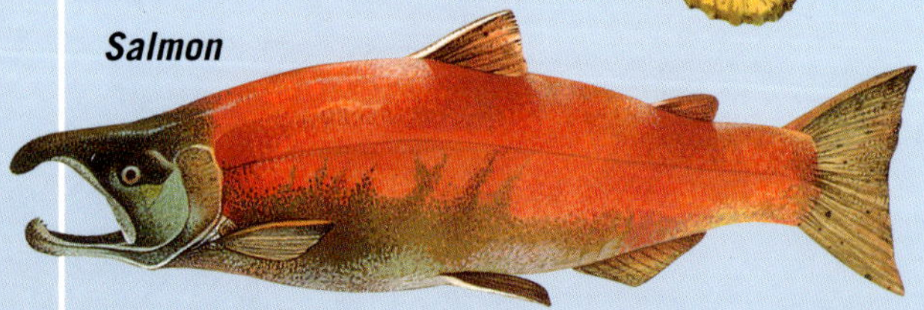

Salmon

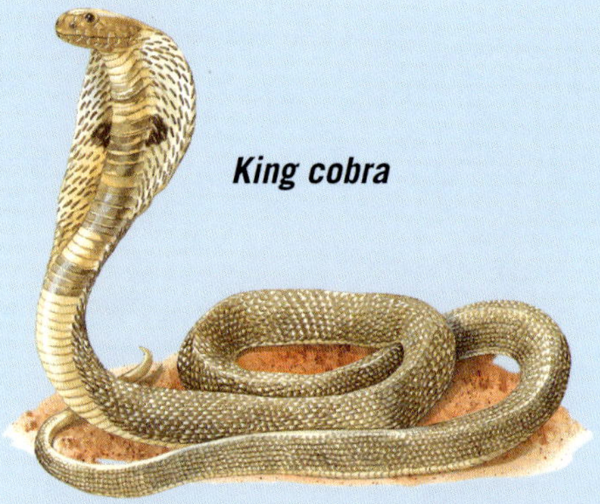

King cobra

Reptiles: cold-blooded, mainly land-living animals with scaly skins. They include lizards, snakes, turtles, and crocodiles. Most reptiles lay eggs. Reptiles need to bask in the sun to raise their body temperature before they can move.

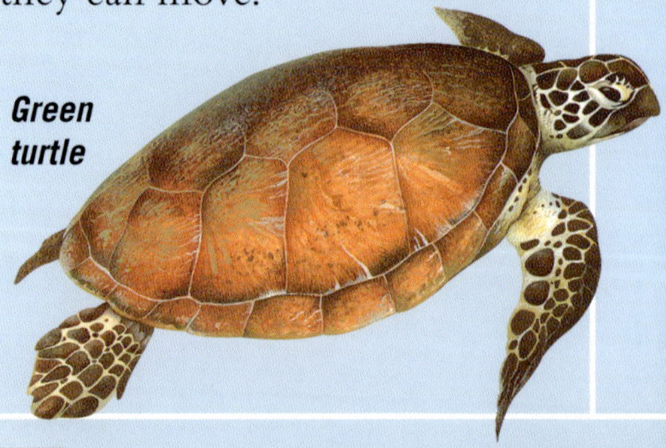

Green turtle

Birds: warm-blooded, feathered animals with front limbs adapted into wings for flight.

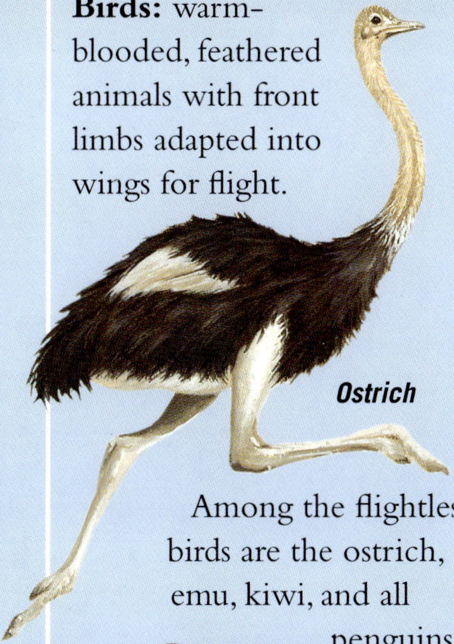

Ostrich

Among the flightless birds are the ostrich, emu, kiwi, and all penguins.

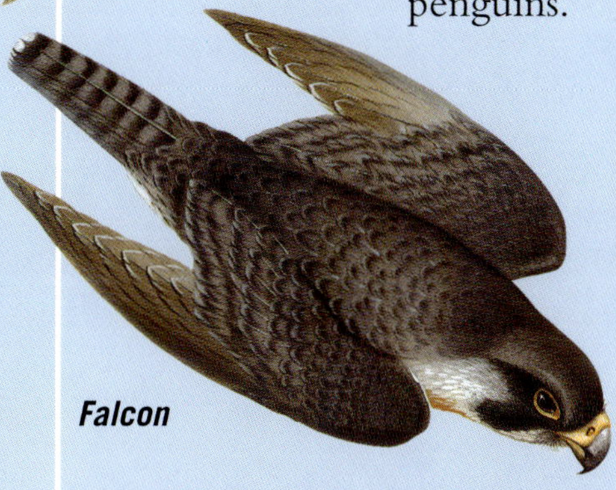

Falcon

There are about 9,000 species of birds. They include seabirds, waterfowl, birds of prey, storks, swifts, kingfishers, cuckoos, and many types of perching bird.

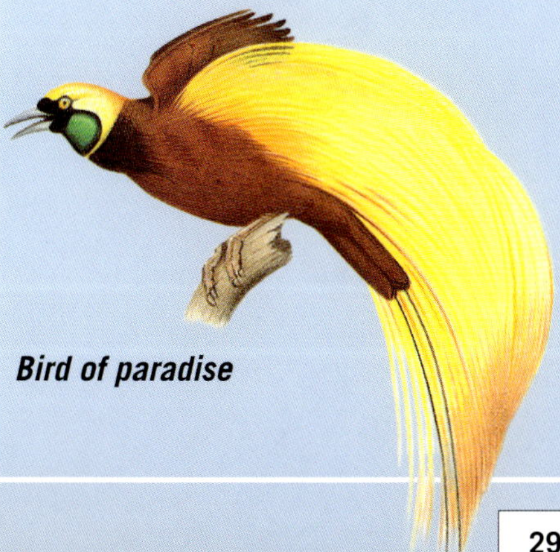

Bird of paradise

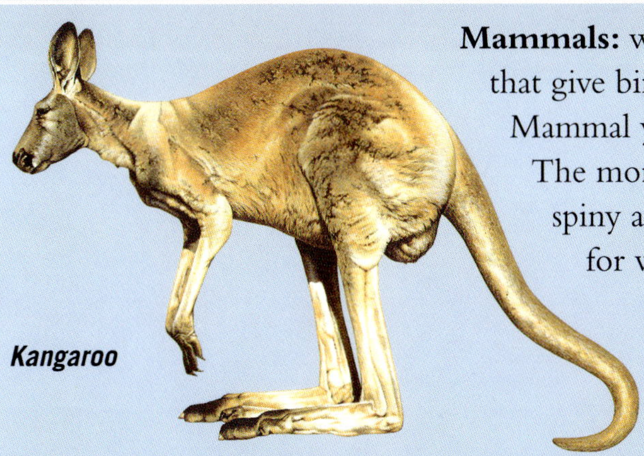

Kangaroo

Mammals: warm-blooded animals that give birth to live young. Mammal young are fed with milk. The monotremes (platypus and spiny anteater) lay eggs. Except for whales and dolphins, all mammals have four limbs and a covering of hair or fur.

Marsupials are mammals that do not give birth to fully developed young. The young are born at an early stage, then develop in their mother's pouch. Marsupials include kangaroos, koalas, and opossums.

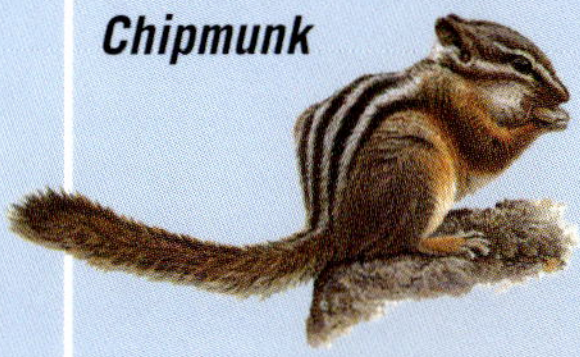

Chipmunk

Mammals have adapted to a wide range of habitats, such as oceans, polar regions, and mountains. Mammals range in size from tiny rodents to giant whales. Some mammals graze or eat leaves, while others are flesh eaters. Bats are the only flying mammals.

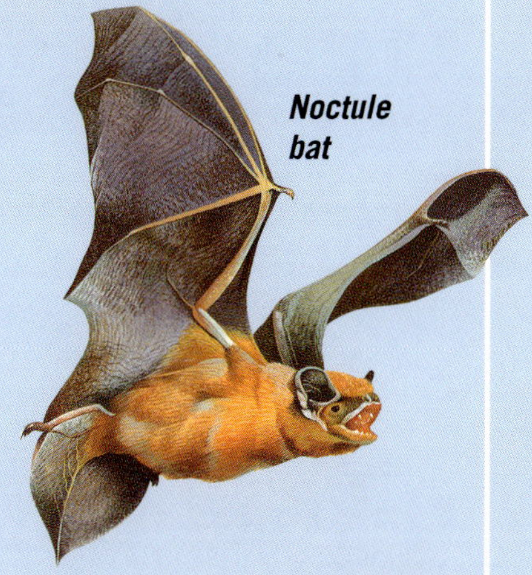

Noctule bat

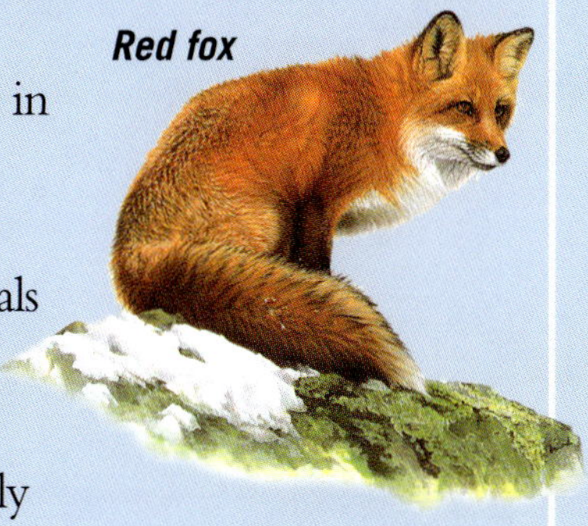

Red fox

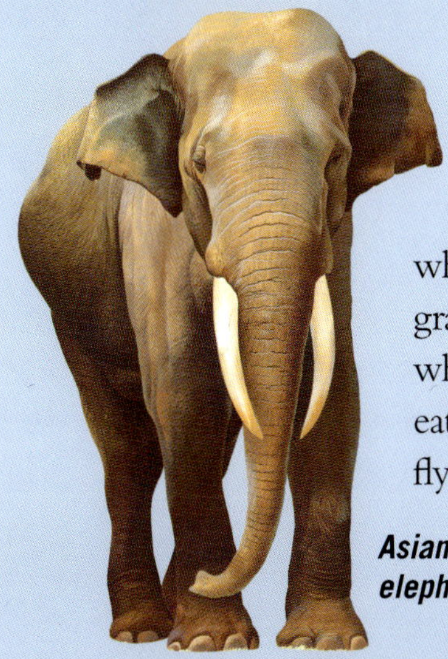

Asian elephant

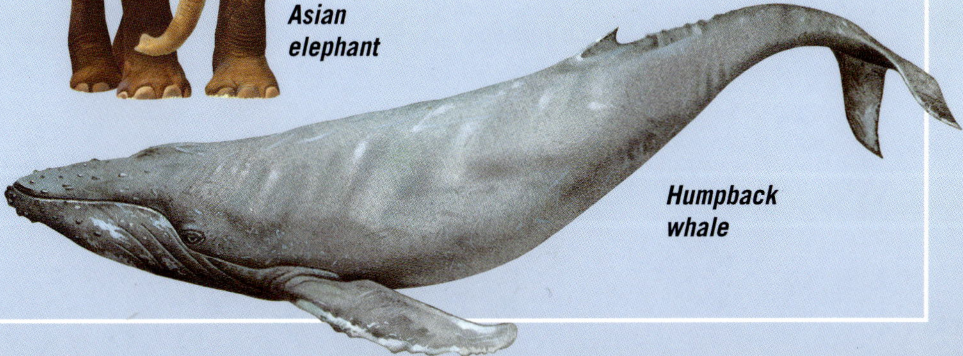

Humpback whale

GLOSSARY

Arthropods Animals with hard external skeletons and pairs of jointed legs. They include insects, crustaceans, spiders, and millipedes.

Atmosphere The envelope of gases that surrounds a planet, moon, or star.

Bacteria Tiny organisms made up of only one cell. Some live inside other organisms. Some types cause disease. Bacteria play a vital role in recycling nutrients in the soil.

Biome A large group of habitats that are generally similar to one another.

Caldera A round basin formed after a violent volcanic eruption during which the volcano collapses in upon itself.

Cell A tiny "building block" that makes up all the tissues in all living things.

Climate The pattern of weather in a particular region of the world over a long period of time.

Crust The thin, rocky outer layer of Earth.

Food chain The sequence in which a plant is eaten by an animal, which is then eaten by another animal, and so on. Because animals eat more than one kind of food, food chains are part of more complex food webs.

Fungi Living things, such as mushrooms, that feed on rotting plant and animal material and reproduce by shedding spores.

Habitat The type of surroundings in which a plant or animal lives.

Invertebrates Animals without backbones.

Magma Hot melted, or molten, rock, that is formed beneath Earth's crust.

Mantle The layer of Earth, which lies beneath the crust and the core, its innermost portion.

Migration The movement of a population of animals from one place to another at a certain time of year to feed or breed.

Moneran An organism that is made up of a single cell without a nucleus. Bacteria are monerans.

Nutrients Substances needed to maintain an organism's bodily activity and new growth.

Organism Any living thing

Photosynthesis The process by which plants use sunlight as an energy source to turn carbon dioxide and water into the sugars they need for food.

Pollen Microscopic grains that are produced in the male part of a flower and transferred to the female part of another (or the same) flower during pollination.

Protist An organism that is made up of a single cell containing a nucleus.

Rift valley A valley formed by the sinking of land between parallel faults.

Savanna Tropical grasslands with scattered trees and bushes.

Tectonic plates The large slabs into which Earth's surface is divided. The plates move relative to one another.

Weathering The action of rain, wind, heat, or frost on decaying or disintegrating rocks.

Vertebrates Animals with backbones.